GALAXY

GALAXY

Collins
An imprint of HarperCollins Publishers
Westerhill Road
Bishopbriggs
Glasgow
G64 8QT

© HarperCollins published in association with Royal Museums Greenwich, the group name for the National Maritime Museum, Royal Observatory Greenwich, Queen's House and Cutty Sark 2013

Maps © CollinsBartholomew Ltd

First published 2013

ISBN 978-0-00-750124-3

Imp 001

The contents of this edition of Collins Galaxy are believed correct at the time of printing. Nevertheless the publishers can accept no responsibility for errors or omissions, changes in the detail given, or for any expense or loss thereby caused.

Printed and bound in China
Design ©HarperCollins Publishers

British Library Cataloguing in Publication Data.
A catalogue record for this book is available from the British Library.

visit our websites at:
www.harpercollins.co.uk
www.collinseducation.com
www.collinsbartholomew.com

GALAXY

EXPLORE THE UNIVERSE, PLANETS AND STARS

Collins

ROYAL
OBSERVATORY
GREENWICH

CONTENTS

FOREWORD

Since human beings first came into existence we have been looking up at the stars and asking questions. Astronomy is a vast and rapidly evolving science in which every new discovery pushes back the limits of our understanding of nature. Whether you are a pupil, student or interested beginner, this book is a great introduction to astronomy. Packed full of the very latest space discoveries with short, simple explanations and diagrams, and a wealth of images taken by a variety of telescopes, Galaxy is easy to follow and builds a firm base for advanced learning. The book takes us on a visual journey through the Solar System, where we learn about our rocky, gassy and icy neighbourhood, and understand what it would be like to visit these places and whether there might be life in other places than Earth. We learn about our active Sun, how it works and how it compares to other stars within our galaxy, the Milky Way. You will also find useful information about how to be an astronomer in your garden or local park, starting with just your eyes and looking further with binoculars or a telescope. The guide to constellations will help you become familiar with the changing night sky and you can learn about the many types of objects you might find: from beautiful star factories to colliding galaxies and our local planets. At the Royal Observatory, Greenwich, we believe the night sky is for everyone and this is the perfect beginner's guide to help you understand our truly awesome Universe better.

Dr Radmila Topalovic
Astronomy Programmes Officer
Royal Observatory Greenwich

THE MILKY WAY

Our Sun is just one of around 200 billion stars in our galaxy, the Milky Way. Look up on a clear night and you'll see a faintly glowing ribbon running across the sky. Our galaxy is named after this 'milky' band, which contains most of those billions of stars. But every star that you can see with the naked eye lies within our galaxy. There may be as many planets orbiting stars as there are stars in the Milky Way.

Until the middle of the twentieth century, it was thought that our galaxy was the only one in the Universe. Then astronomer Edwin Hubble proved that the Universe is sprinkled with galaxies. There are around 200 billion other galaxies, each with billions of stars of its own. There are more stars in the Universe than there are grains of sand on all the beaches on Earth.

Galaxies come in different shapes and sizes. The Milky Way looks like a ribbon from Earth, but it is actually shaped like a very thin pancake with a bulge in the middle. Just as Earth is not at the centre of our Solar System, neither is the Solar System at the middle of our galaxy. It lies about two-thirds of the way out from the centre of this galactic pancake. Since we are looking at our thin galaxy from the side, it appears as a strip.

However, if you could see the Milky Way from above you would notice a bar-shaped core surrounded by a disk of gas, dust and stars made from two swirling spiral arms. This shape is called a 'barred spiral' galaxy. At the centre of our galaxy is a very massive and dense object, probably a black hole. The stars, dust and gas in our galaxy rotate around this centre. Our solar system takes about 225 million years to make one rotation.

Distances in space are so huge that they are measured in light-years – the distance a beam of light would travel in one year. Light travels incredibly quickly (300,000 kilometres a second) but still takes 8 minutes to reach Earth from our Sun. Light would take 120,000 years to cross the Milky Way. To visualize this vast distance, imagine our Solar System as far out as Pluto was shrunk to the size of a 10 pence coin (25 mm). The Milky Way would then be a disc 2,000 km across – which would stretch from London to Reykjavik in Iceland.

Curiously, although the Milky Way shines with the light of billions of stars, it is mostly empty space. If you shot our Sun through the disk like a snooker ball, the chances of it hitting another star before zooming out the other side would be one in a billion. That's seventy times less likely than winning the lottery!

THE MILKY WAY

All the stars that we can see with the naked eye in the night sky are part of our home galaxy, the Milky Way. It is a 'barred spiral' galaxy, similar in shape to a spinning firework. Our Solar System lies about two-thirds of the way out from the centre of it on the spiral-shaped arm of Orion. The highest density of stars is near the galactic centre, drawn together by gravity where the arms converge, and it is these stars that we can sometimes see as a hazy band of white light stretching across the sky on a clear night. Nearer stars appear as sharper points of light. Dark clouds of dust and gas, called nebulae, obscure some stars and other star-forming nebulae shine with reflected light or produce their own light if heated to very high temperatures.

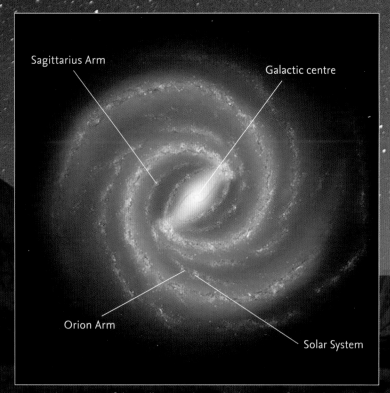

Sagittarius Arm

Galactic centre

Orion Arm

Solar System

An artist's concept of the Milky Way shows the location of our Solar System within its galaxy. Since it was formed our Sun has rotated sixteen times around the galaxy

M24, the Sagittarius Star Cloud, is a bright streak of densely concentrated stars in the Sagittarius Arm of the Milky Way

M8, the Lagoon Nebula, is a bright star-forming cloud between 4,000 and 6,000 light-years from Earth

The Pipe Nebula, a large pipe-shaped cloud of dust and gases that obscures the stars behind it

Antares, a red supergiant star

Galactic centre

M6 is a butterfly-shaped cluster of stars about 100 million years old and about 2,000 light-years from Earth

It is impossible to take one photograph of all of the Milky Way but this image taken in the semi-desert of Bardenas in northern Spain looks towards the centre of our galaxy, about 27,000 light years away

THE SOLAR SYSTEM

The Solar System is the Sun and the many objects that orbit it. These objects include eight planets, at least five dwarf planets and countless asteroids, meteoroids and comets. Orbiting some of the planets and dwarf planets are over 160 moons. The Sun keeps its surrounding objects in its orbit by its pull of gravity, which has an influence for many millions of kilometres. The Solar System lies within one of the outer arms of Milky Way galaxy, which contains about 200 billion other stars.

Around 4.6 billion years ago the Solar System was a giant cloud of dust and gas. This then collapsed under its own gravity, possibly as a result of a shockwave from a nearby exploding star. Nearly all of the cloud's material went into forming the Sun, but there was some stuff from the nebula left over that didn't make it. This material formed a flattish disc around the young star. The heavier lumps of rock and metal drifted closer in towards the Sun. These then clumped together into bigger chunks, eventually becoming the rocky inner planets, Mercury, Venus, Earth and Mars.

The outer planets, Jupiter, Saturn, Uranus and Neptune, began in the same way but their cores got so big they captured leftover gas from the nebula. They grew into gas giants rather than rocky planets. The biggest planets, Jupiter and Saturn, are mainly composed of hydrogen and helium. The two outermost planets, Uranus and Neptune, also contain ices of water, ammonia and methane. Rings of dust and other particles encircle each of the outer planets.

There are five individual objects big enough to have been rounded by their own gravity: Ceres, Pluto, Haumea, Makemake and Eris.

They are called dwarf planets. Astronomers believe many more dwarf planets will be discovered.

The planets orbit the Sun in paths that are almost circular and in a nearly flat disc called the ecliptic plane. Six of the planets and three of the dwarf planets are orbited by natural satellites, called 'moons' after Earth's Moon.

The Solar System is also home to regions of smaller objects. The asteroid belt between Mars and Jupiter is an area of countless rocky bodies. Beyond Neptune's orbit lies the doughnut-shaped Kuiper belt, which is similar to the asteroid belt but 20 times as wide. Pluto is the largest object in the Kuiper belt. Further out still is the scattered disc of icy bodies, many of which move inwards to become comets.

The solar wind is a flow of plasma that expands out from the Sun in every direction. It created a kind of bubble in space known as the heliosphere, which extends out to the edge of the scattered disc. A thousand times further out still, at the deepest darkest reaches of the Solar System, is a huge shell of billions of icy lumps called the Oort cloud.

THE SOLAR SYSTEM

About 500 years ago most people still believed that the Earth was the centre of the Universe. Our five nearest planets are visible to the naked eye so mankind has been able to observe them since his earliest time on Earth, but the idea that they orbited the Sun, and not the Earth, was still controversial when published by Copernicus in 1543. Knowledge of our Solar System expanded with the development of telescopes in the seventeenth century. Over time it was found that other planets had moons as well as our own and that there were further planets beyond Saturn. It was also realized that the Solar System contains countless asteroids and comets, all debris from the birth of a star and the formation of our planetary system billions of years ago.

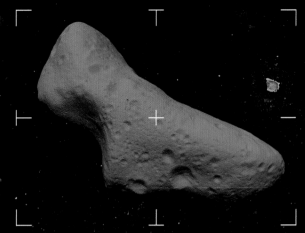

34 km-long Eros was the first asteroid to be orbited by a robotic space probe. After a five-year voyage the probe touched down in 2001, sending close-up images back to Earth

The Sun and its eight planets

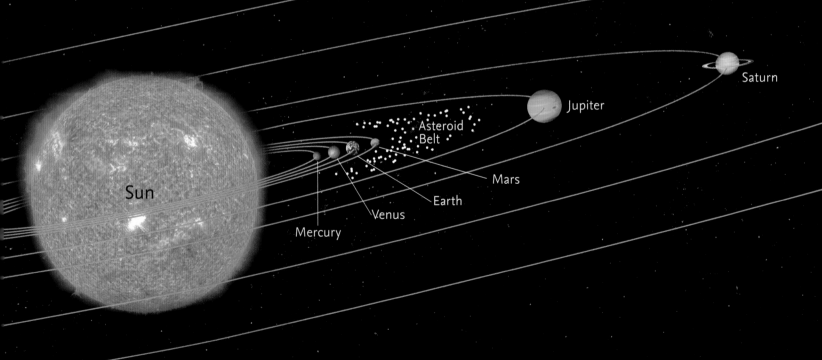

Sun

Mercury

Venus

Earth

Mars

Asteroid Belt

Jupiter

Saturn

A meteoroid heats up when it enters the Earth's atmosphere to become a light-emitting fireball called a meteor or shooting star

Neptune

Uranus

Comet

The comet Hale-Bopp was visible from Earth in 1997 and was hailed as the most-observed comet of the twentieth century. It will next return to the inner Solar System in about the year 4385

THE SUN

The Sun is the star of our Solar System. If you could get close, you'd find it a pretty unfriendly place. This gigantic sphere of superhot glowing gas is undergoing constant nuclear fusion and pumping out fierce magnetic fields. But it's only thanks to this cosmic inferno that we exist. Heat and light from this star travel millions of kilometres to support all life on our planet. Sunlight is Earth's primary source of energy. It warms us during the day, powers our weather systems and plants capture its energy through photosynthesis. Even the energy in oil was captured by photosynthesis in the past.

The Sun is vast: it contains 99 per cent of all matter in the Solar System and weighs 700 times more than all the other planets combined. You could fit 1 million Earths into it. This huge mass means its gravity pulls on everything in the Solar System. If the Earth were the size of a cherry, the Sun would be a ball 1.5 m across (about the height of an adult person). The cherry would orbit the ball at a distance of 150 m, the length of one and a half football pitches. Three quarters of the Sun's mass is hydrogen, the rest is mostly helium with smaller amounts of heavier elements including oxygen, carbon, neon and iron.

Our sun was born 4.6 billion years ago from a huge cloud of gas and dust called a nebula, which collapsed in on itself under the pull of gravity. This became denser as it heated up, with its core reaching a scorching 14 million °C. At this temperature nuclear fusion started and the Sun became a star and shone. The surface is a relatively cool 5,500 °C. This is the part of the Sun that we see, and is called the photosphere.

The corona is the Sun's outer atmosphere, which expands continuously into space throughout the Solar System. The corona is much hotter than the actual surface of the Sun. It can be clearly seen in a solar eclipse when the Moon covers the main body of the Sun (but you must never look directly at the Sun). Sunspots are darker, cooler splotches on the photosphere that signify areas of intense magnetic activity. The largest sunspots can be tens of thousands of kilometres across. The areas around sunspots are the source of solar flares and coronal mass ejections, events where huge quantities of matter and radiation are shot from the sun into space. These can stretch for many thousands of kilometres and would dwarf the Earth.

The Sun's core is a vast nuclear power station, fusing hydrogen fuel into helium and releasing energy as it does so. A hydrogen bomb – the most powerful device created by man – releases the equivalent of 10 million tons of TNT on detonation. The Sun releases 10 billion times this amount of energy every single second. Energy works its way from the nuclear reactor core to the outside and radiates across space. The Sun fuses 620 million tons of hydrogen every second. It has plenty of fuel and will still be burning 5 billion years from now.

THE SUN — OUR STAR

A huge ball of scorching hot glowing gases, the Sun is large
enough to contain every single other object in the Solar System.
It is almost ten times wider than the next largest object, which is
Jupiter. Heat and light from this star travels millions of kilometres
to reach Earth, taking about eight minutes to reach it and another
five hours to reach the outer fringes of the Solar System.
Already burning for 4.6 billion years, the Sun is still only about
halfway through its life span. Yet vast though it is, the Sun is only
average in size and brightness compared to other stars. From the
far reaches of space it is a yellow dwarf, a quite ordinary yellow-
white star of only medium temperature compared to many others.

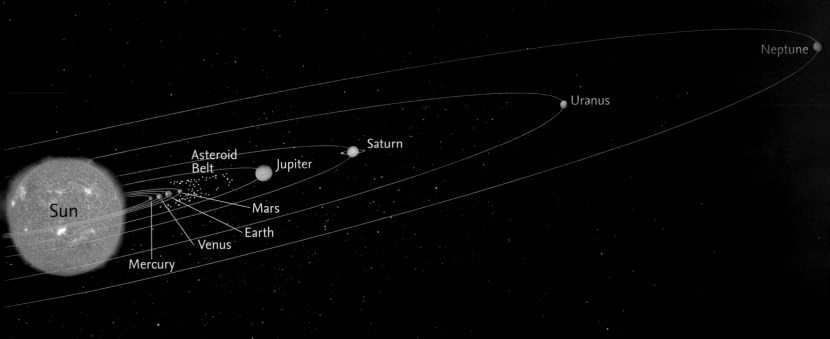

Sun

Mercury

Venus

Earth

Mars

Asteroid
Belt

Jupiter

Saturn

Uranus

Neptune

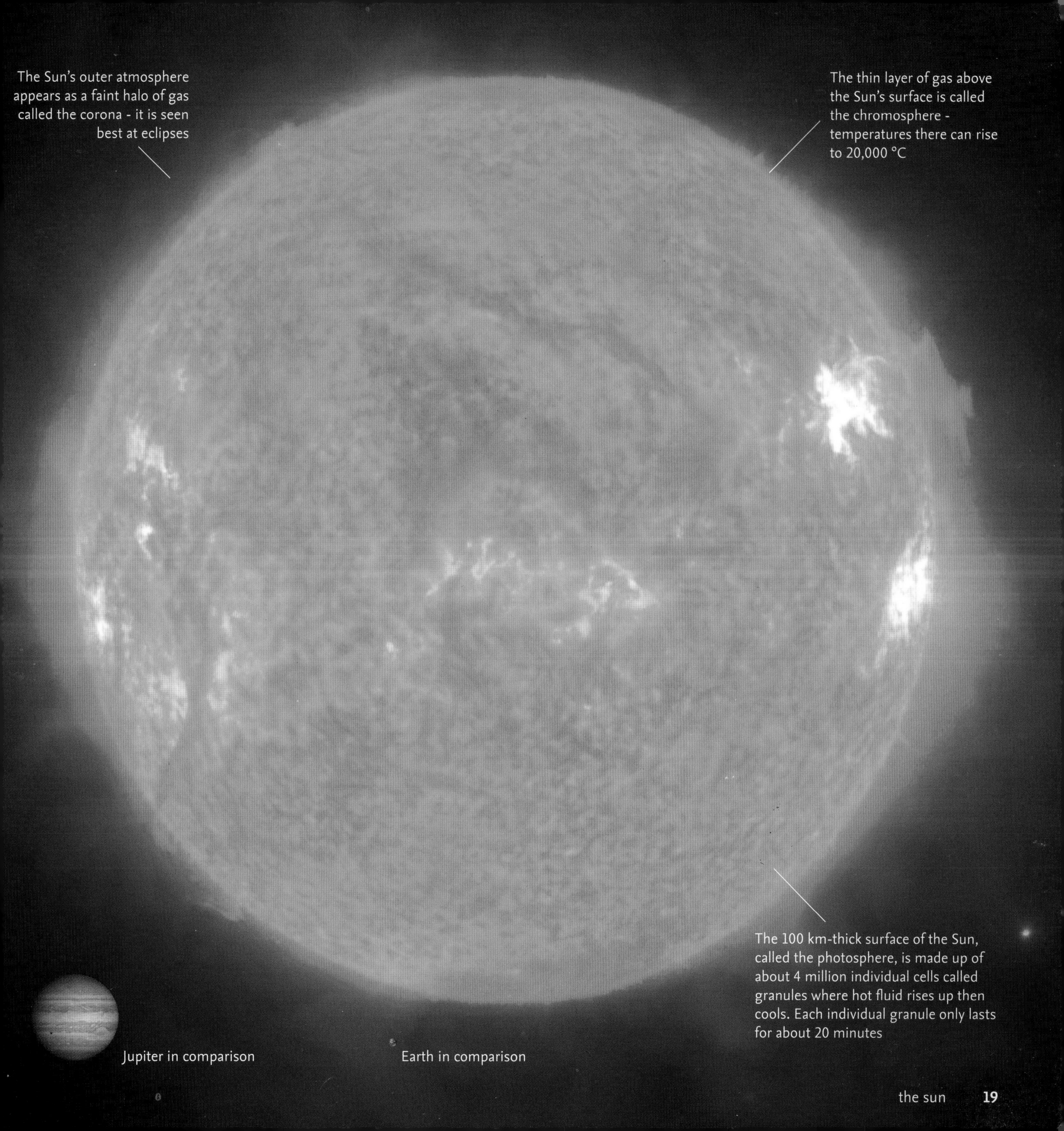

The Sun's outer atmosphere appears as a faint halo of gas called the corona - it is seen best at eclipses

The thin layer of gas above the Sun's surface is called the chromosphere - temperatures there can rise to 20,000 °C

The 100 km-thick surface of the Sun, called the photosphere, is made up of about 4 million individual cells called granules where hot fluid rises up then cools. Each individual granule only lasts for about 20 minutes

Jupiter in comparison

Earth in comparison

THE SUN — FACTS

150 million km

Distance from Earth

230 million Earth years

Length of galactic year

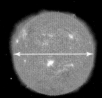

25 Earth days 9 hours

Length of day

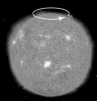

1 391 016 km

Diameter

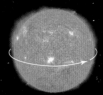

4 370 000 km

Circumference

5504 °C

Average temperature

4.6 billion years

Age

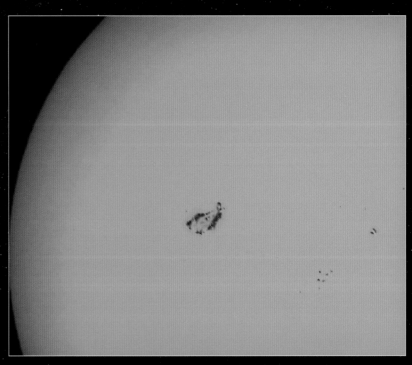

Temporary areas of intense magnetic activity cause dark sunspots on the Sun's surface. At less than 4,000 °C they are cooler than the rest of the photosphere. Appearing in many shapes and sizes, and often in groups, they can be over ten times the diameter of Earth and can last between several days and several weeks

Earth in comparison

A massive solar flare, or prominence, erupts from the Sun's
surface, releasing as much energy as millions of hydrogen
bombs in just a few minutes.

THE INNER SOLAR SYSTEM

Mercury, Venus, Earth and Mars are known as the inner planets because they are closest to the Sun. They are also called the terrestrial planets because their surfaces are solid and rocky. Also in the inner solar system is the asteroid belt, a region beyond the orbit of Mars where thousands of asteroids orbit the Sun. Beyond that is the outer solar system where the gas giants Jupiter, Saturn, Uranus and Neptune orbit.

The terrestrial planets differ from the gas giants in a number of ways. They are much smaller and have few or no moons. Earth has one moon, Mars has two, but Mercury and Venus have none at all. Jupiter has 67 moons! Nor do the terrestrial planets have planetary rings, as the gas planets do.

Terrestrial planets have similar basic structures. At the heart of each is a core made from heavy metals, mostly iron. This core is surrounded by a mantle of silicate rock. On the surface are features that we would find familiar: mountains, volcanoes, canyons and craters.

But the inner planets are very different from each other when it comes to atmospheres. Venus has a thick toxic atmosphere of carbon dioxide, which traps the Sun's heat making it the hottest planet in the Solar System. It is nearly the same size as Earth. Mercury, the smallest terrestrial planet, has almost no atmosphere at all. The heat its daytime side gains is lost at night, so it alternates between burning and freezing temperatures. Mercury is dense, being made mostly of iron and nickel.

Earth's atmosphere is mostly oxygen, nitrogen, and carbon dioxide with traces of other substances. The Earth itself has a high percentage of iron, making it the densest of all the planets. Mars is the fourth terrestrial planet in the Solar System. Its rusty red appearance is just that – the iron in its surface rocks has rusted! It has a thin atmosphere, but scientists believe that it was once thicker. When the atmosphere thinned any liquid water Mars had evaporated.

Beyond Mars lies the asteroid belt, an area occupied by thousands of irregularly shaped bodies called asteroids or minor planets. These are mostly unused building blocks of the material that formed the planets billions of years ago. Most asteroids are only a kilometre or so across. But Ceres, the largest, is 950 km in diameter and is considered a terrestrial dwarf planet, the only one in the inner solar system. It has a rocky inner core with an icy mantle and varied terrain like the other terrestrial planets. Around a quarter of the planet is made up of water ice. Ceres might also have a thin atmosphere. The remaining bodies vary in size right down to dust particles. Although called a 'belt', the asteroid material is so spread out that several unmanned spacecraft have passed through without any hitting anything.

THE INNER SOLAR SYSTEM

Beginning as material left over from the formation of the Sun, all four inner planets gradually formed into balls of hot rock in a process that took 100 million years. Known as the terrestrial planets, they share a similar structure with a solid crust surface of rock that was once molten before it cooled and solidified. Over hundreds of millions of years these rocky surfaces were bombarded with other lumps of rock from space. On Mercury and the Moon the impact craters can still be seen clearly although on Earth most of them have been obscured by volcanic activity or weathering. Since formation the inner planets have evolved very differently – the result of different chemical composition, size and distance from the Sun – but the fact remains that although atmospheric probes may investigate the outer planets it is only on the inner ones that a physical landing is possible.

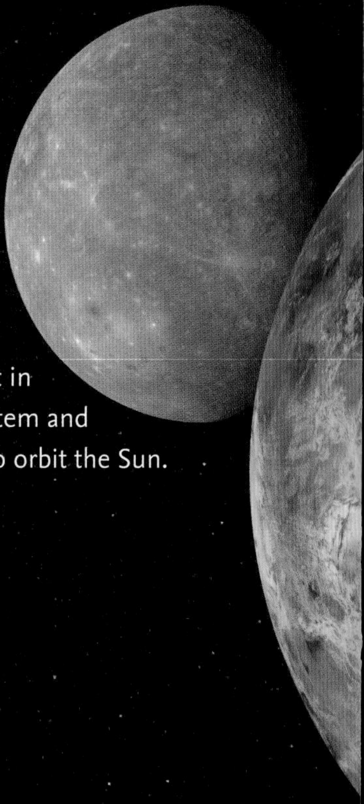

Mercury
The smallest planet in the Inner Solar System and the fastest planet to orbit the Sun.

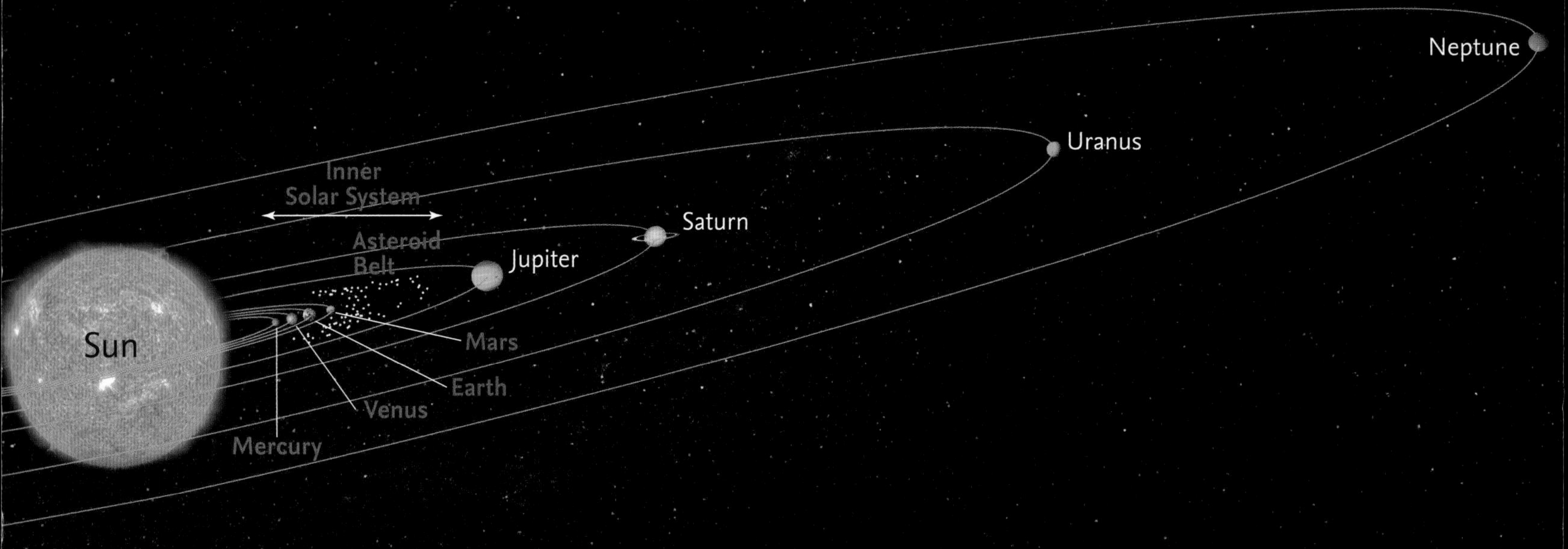

Neptune

Uranus

Inner Solar System

Asteroid Belt

Saturn

Jupiter

Sun

Mars

Earth

Venus

Mercury

Earth
The only planet in
the Solar System
known to have life.

Venus
The hottest planet in the Solar
System is cloaked in clouds of
sulphuric acid that prevent its
surface being seen from Earth.

Ceres
A nearly spherical
dwarf planet, and the largest
object in the Asteroid Belt.

Mars
Named after the
Roman god of war,
it is often described
as the "Red Planet".

MERCURY

Mercury is the smallest planet and is not much bigger than the Moon. It is closest to the Sun and takes only eighty-eight Earth days to complete an orbit of the Sun. It zooms round faster than any other planet. Mercury is dotted with innumerable craters, which make it also look a lot like our Moon. The craters were made by comets and asteroids crashing into the planet billions of years ago, when large amounts of debris were flying around the inner solar system. The planet is named after Mercury, the messenger to the Roman gods.

Visiting Mercury wouldn't be a lot of fun. There is no air to breathe and no water. Even if you had a spacesuit to supply these, you would still need thick insulation and shielding to handle the planet's wild temperature swings. During the day Mercury's surface reaches a scorching 400°C (hotter than your oven),

while at night it would fall to -185°C (far colder than even the chilliest part of Antarctica). This huge variation is because the planet has no atmosphere to retain its heat.

At least walking about would be a bit easier as the planet's gravity is just a third of Earth's. While exploring, you would traverse a rocky, crater-strewn land where huge cliffs shoot up from the barren plains. The Caloris Basin is one of the largest of Mercury's surface features. It was created when a massive asteroid hit the planet 3.8 billion years ago. The round basin is 1,550 km across and the mountains in its walls are 2 km high. The floor is filled with ancient lava plains.

Mercury isn't tilted on its axis like Earth. So if you travelled to its north or south poles, the Sun would barely rise above the horizon.

This also means that the floors of deep craters at the poles are permanently in shadow and so are always very cold. In 2012 NASA's space probe Mercury MESSENGER confirmed that craters at the north pole contained large deposits of water ice. This was most likely deposited by icy comets smashing into the craters.

Another big difference you would notice is the size of the Sun. It would appear three times larger in the sky than on Earth, because it is so much closer to Mercury.

It is tricky to see Mercury from Earth with the naked eye because it is always close to the glaring sphere of the Sun. But if you look near the horizon just before sunrise or after sunset you can sometimes see it shining like a bright star.

MERCURY –
THE CLOSEST PLANET TO THE SUN

A dry and airless place, Mercury has scorching hot temperatures during the day and freezing ones at night. It is the least explored of the inner planets. In the 1970s about 45 per cent of the planet's surface was photographed by Mariner 10. Then in 2011 NASA's MESSENGER became the first spacecraft to orbit Mercury, relaying back images of the planet that had never been seen before.

Looking a lot like the Moon, the images show a surface dominated by craters of all shapes and sizes. Craters range in size from under 100 metres in diameter to the giant Caloris Basin, large enough to contain all of England. With no wind erosion, flowing water or recent volcanic activity to obscure these craters this means that Mercury has the oldest land surface in the Solar System.

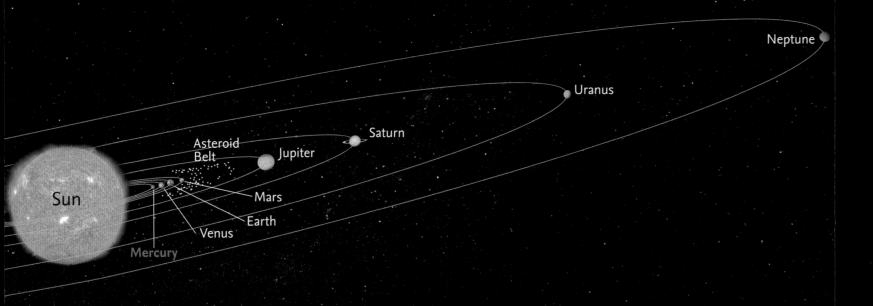

Neptune

Uranus

Saturn

Asteroid
Belt   Jupiter

Sun

Mars

Earth

Venus

Mercury

MERCURY — FACTS

58 million km

Distance from Sun

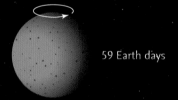

88 Earth days

Length of year

59 Earth days

Length of day

4900 km

Diameter

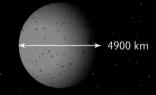

15 300 km

-173 °C to 427 °C

Average temperature

None

Number of moons

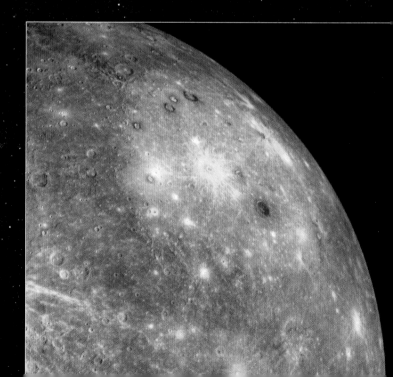

Van Eyck

Sophocles

Vivaldi

Tolstoy

Michelangelo

Bach

Craters on Mercury are named after famous writers, musicians and authors who have been dead for at least three years. Names are approved by the International Astronomical Union (IAU). In 2012 nine craters in the northern polar region were named, including a 48 km-wide one called Tolkein after J.R.R. Tolkein, author of *The Hobbit* and *The Lord of the Rings*. Later the same year a Mickey Mouse-shaped crater in the southern hemisphere was named after the filmmaker and animator Walt Disney

On the rare occasions that Mercury passes directly between the Earth and the Sun it can be seen as a tiny black dot moving across the face of the Sun. This happens about thirteen times a century, in either May or November, with the next occurrence in May 2016

Sun

VENUS

Venus is named after the Roman goddess of love because of its brightness and beauty. It is the brightest object in the night sky after Earth's moon. It is the second planet from the Sun in the Solar System and the nearest planet to Earth. Venus is a similar size to Earth and it is sometimes called our sister planet. The diameter of Venus is 12,092 km (only 650 km less than the Earth's) and its mass is 81.5 per cent of the Earth's.

But its beauty belies its true nature: if Hell is anywhere in the Solar System it is on Venus. Its surface temperature of 480°C is hot enough to melt lead, making it the hottest planet by far. It is totally cloaked in a thick, poisonous atmosphere of carbon dioxide with some nitrogen and topped by clouds of sulphuric acid. These dense clouds act as a huge greenhouse, trapping heat. Imagine sitting in a car on a summer day: sunlight blazes through the windows warming you and the seats, but this heat cannot escape. Venus has been stuck in its 'hot car' for billions of years!

The atmospheric pressure on the surface is 92 times that on Earth's surface and the same as at 1,000 m down in Earth's oceans. If you stepped onto the surface you would be crushed within seconds. Violent thunderstorms with lightning flashes rage through the atmosphere. The cloud tops are whipped by 300 km/h winds and produce sulphuric acid rain. Dense clouds mean the Sun would be a dull orange smudge and you would never see any stars.

This hellish atmosphere makes observing Venus's surface tricky. Several spacecraft have made it to the Venusian surface and have sent back images. But they only lasted a couple of hours in the fierce conditions. Radar surveys by the orbiting Magellan spacecraft have produced clearer images.

Venus was shaped by volcanic activity. About 80 per cent of its surface is covered by dusty volcanic plains dotted with slab-like rocks. There are also two highland areas with some mountains. The tallest peak is Maxwell Montes, which stands 11 km high, topping Mount Everest by 2,000 m. Venus has several times as many volcanoes as Earth, including hundreds that are more than 100 km across. Oceans may once have lapped Venusian shores, but these were vaporized as the temperature rose due to a runaway greenhouse effect.

Seen from Earth, Venus reaches its maximum brightness shortly before sunrise or shortly after sunset, for which reason it has been referred to by ancient cultures as the Morning Star or Evening Star. Venus is the brightest point-like object in the sky and is often misreported as a UFO!

VENUS — EARTH'S NEIGHBOUR

Venus is Earth's nearest neighbour but although it is similar in structure and size it is a very different planet. Venus is a volcanic world, perpetually cloaked in clouds that form a thick, rapidly swirling atmosphere. Temperatures are hotter than on any other planet in the Solar System. It has a retrograde spin, turning in an opposite direction to the other planets in the Solar System. This means that it is the only planet where the Sun rises in the west and sets in the east. It also spins very slowly and is the only planet that takes longer to spin on its axis than it does to orbit the Sun – meaning that a Venus day is actually longer than a Venus year.

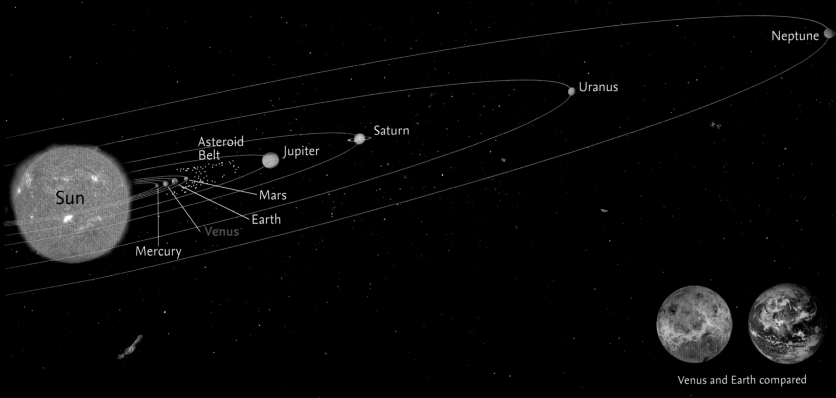

Venus and Earth compared

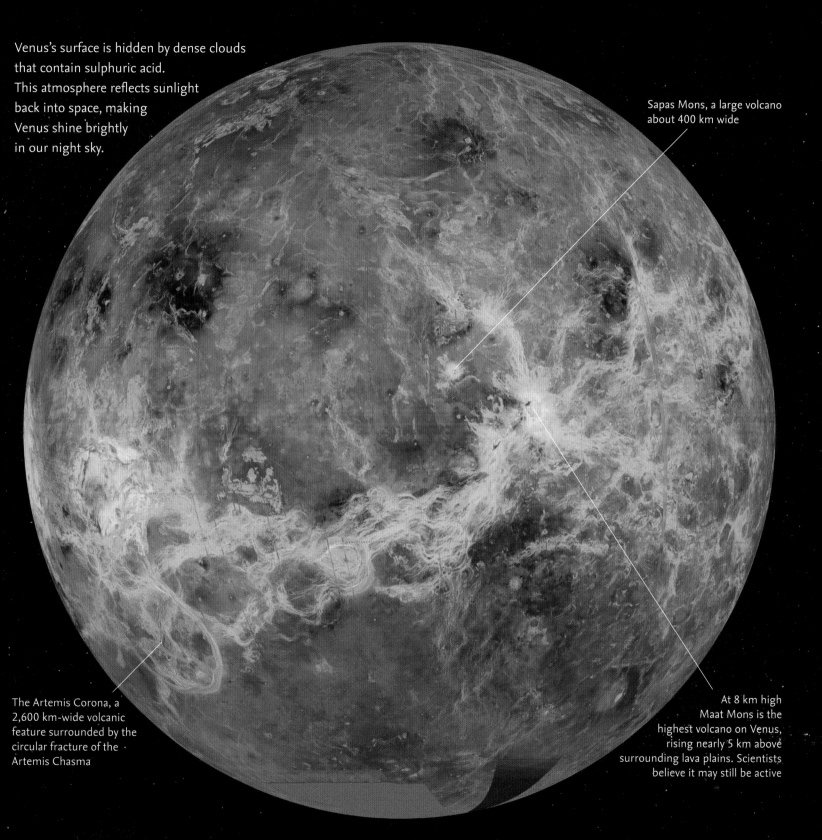

Venus's surface is hidden by dense clouds that contain sulphuric acid. This atmosphere reflects sunlight back into space, making Venus shine brightly in our night sky.

Sapas Mons, a large volcano about 400 km wide

The Artemis Corona, a 2,600 km-wide volcanic feature surrounded by the circular fracture of the Artemis Chasma

At 8 km high Maat Mons is the highest volcano on Venus, rising nearly 5 km above surrounding lava plains. Scientists believe it may still be active

VENUS — FACTS

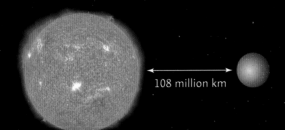

108 million km

Distance from Sun

224 Earth days 17 hours

Length of year

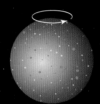

243 Earth days

Length of day

12 100 km

Diameter

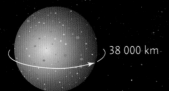

38 000 km

Circumference

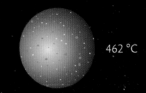

462 °C

Average temperature

None

Number of moons

A false-colour image captured by the Mariner 10 spacecraft in 1974 shows Venus's clouds of sulphuric acid spiralling from the planet's equator to its poles

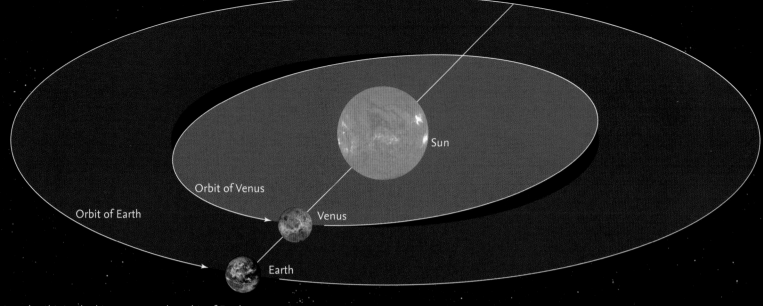

Sun

Orbit of Venus

Orbit of Earth

Venus

Earth

Venus's orbit is tilted in respect to the orbit of Earth

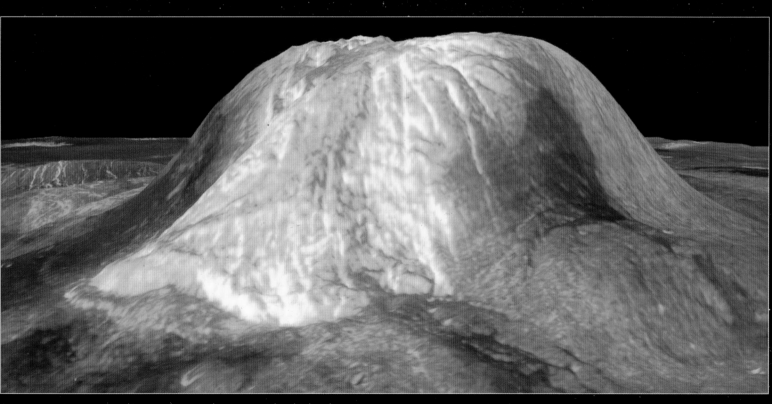

A 3-D computer-generated image shows Gula Mons, a 3 km-high volcano in a part of Venus called Eistla Regio. The volcano is surrounded by lava flows that are hundreds of kilometres long

EARTH

Earth is the third planet from the Sun and the largest of the four rocky inner planets. Oceans, at least 4 km deep, cover 71 per cent of Earth's surface. It takes Earth 365 days and 6 hours (one year) to orbit the Sun. It has one natural satellite, the Moon, which is the only other place to be visited by people from Earth. Thousands of artificial satellites orbit the planet. These help us communicate, chart the weather and watch the stars. Our planet is home to billions of humans and millions of other animal and plant species. Not a single living organism has been found anywhere else in our Solar System. Yet...

Earth formed 4.6 billion years ago in tandem with the Sun. Gravity pulled lumps of rocky material in the disc around the Sun together, creating the young planet. The young planet was drawn together by gravity and its core got very hot, melting the rocks.

A crust formed when the molten outer layer of the planet Earth cooled. The atmosphere and oceans partly came from trapped gases in the mantle, including water vapour that bubbled out during volcanic activity, but most of the water that makes up today's oceans may have arrived here on asteroids and some icy comets.

Life appeared within 1 billion years. Microscopic life in the early oceans converted gases into the air we breathe today and there is more oxygen in Earth's atmosphere today than when the planet first formed. Today the atmosphere is 78 per cent nitrogen and 21 per cent oxygen with small amounts of other gases.

Earth's crust is divided into several rigid segments, or tectonic plates. These float over the mantle over periods of many millions of years. Earth's poles are mostly covered with ice. A solid ice sheet up to 4 km thick covers the southern continent of Antarctica; at the north pole is packed sea ice. The planet's interior remains active, with a solid iron inner core, a liquid outer core that generates the magnetic field, and a thick layer of relatively solid mantle.

The shape of the Earth is a sphere that bulges around the equator. For this reason the furthest point on the surface from the Earth's centre is not Mount Everest but the Chimborazo volcano in Ecuador. Earth's average diameter is about 12,742 km. Although it seems mountainous to us, the Earth is comparatively, smoother than a snooker ball.

Earth is our ideal environment. It has the right amount of gravity to hold us down, it has oxygen in the air to breathe and a Sun to warm us. There is food to eat, natural wonders to observe and amazing places to visit. Weather continually moves the sky above us. There are violent storms, lightning and hurricanes but also warm summers and rich harvests. Earth's atmosphere protects us from harmful radiation and yet allows light to filter through. We can see stars and galaxies billions of kilometres away.

The Sun is slowly becoming more luminous and in 1.5 billion years this will cause Earth's oceans to boil away. In about 5 billion years the Sun will become a red giant. Earth will enter its atmosphere and be vaporized. After that, the Sun's core will collapse into a white dwarf, and the matter that once made up the Earth, human beings and even this book will be ejected into interstellar space. There it will eventually become part of a new generation of stars and planets.

EARTH — OUR HOME IN SPACE

From 300 km above the Earth the major landforms of the planet's surface, and even the lights of cities, are clearly visible. From 300,000 km away the Earth still appears as a bright ball, appearing similar in size as the Moon does to us from Earth. From the edge of the Solar System it is a tiny blue dot. Yet this tiny blue dot is the only place we know of in the Universe that supports life. It is a living planet, with plenty of water, plants and breathable air,

protected by its atmosphere. Most of the 4 billion-year-old meteor impact craters that once dotted its surface have long since eroded and it has become a planet of immense variety in both landforms and life forms. It has taken millions of years to transform Earth into the planet we know and as processes continue, even without human intervention, in millions of years it will have transformed again into a planet that we today would not even recognize.

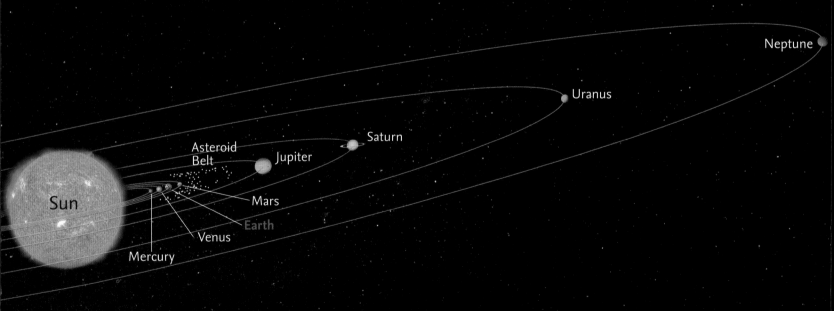

Neptune

Uranus

Saturn

Asteroid
Belt   Jupiter

Sun

Mars

Venus   Earth

Mercury

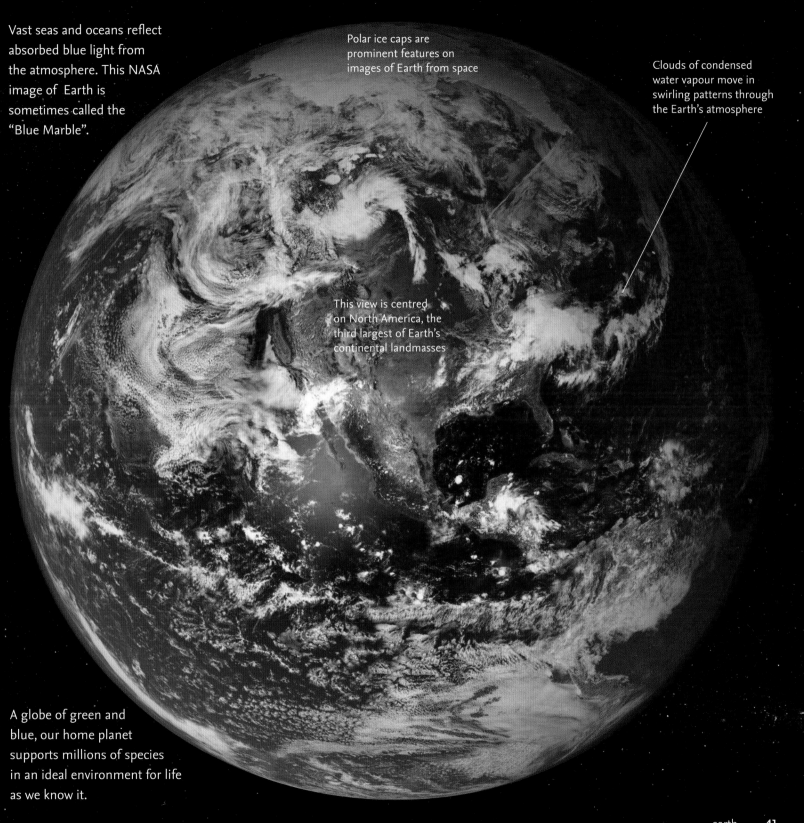

Vast seas and oceans reflect absorbed blue light from the atmosphere. This NASA image of Earth is sometimes called the "Blue Marble".

Polar ice caps are prominent features on images of Earth from space

Clouds of condensed water vapour move in swirling patterns through the Earth's atmosphere

This view is centred on North America, the third largest of Earth's continental landmasses

A globe of green and blue, our home planet supports millions of species in an ideal environment for life as we know it.

ACTIVE EARTH

The Earth is made up of three main layers called the crust, mantle and core. The crust, the outermost layer, is rigid and very thin compared with the other two. Beneath the oceans, the crust varies little in thickness, generally extending to about 5 km. The thickness of the crust beneath continents is much more variable but averages about 30 km and under large mountain ranges, such as the Alps or the Sierra Nevada, the base of the crust can be as deep as 100 km. The Earth's crust is brittle and can break. It is made of separate pieces called plates. The plate boundaries can be violent places where earthquakes and volcanoes occur.

Below the crust is the mantle. This is a dense, hot layer of semi-solid rock 2,900 km thick. The mantle is hotter in some areas and colder in others, and so it moves in a state of convection. You can think of the plates of the crust floating on this planet-wide sea of flowing rock.

The way convection causes the plates of the crust to move is called plate tectonics. In a way, plate tectonics is how the Earth is resurfaced. Where plates are moving apart, new crust is created by the flow of magma to the surface. This is called seafloor spreading. When one plate is pushed beneath another it is called subduction. Often the plates stick and then jolt violently to release the pressure, causing earthquakes.

Volcanoes are usually caused by subducted crust material being melted in the magma and then forcing its way back to the surface. The Himalaya Mountains, the world's tallest mountain range, were formed by the collision of two major plates. Before being uplifted they were covered by an ancient ocean. Plate tectonics may occur elsewhere in the Solar System.

At the centre of the Earth lies the core, the inner part of which is solid. The semi-liquid outer core flows in continuous currents around the inner iron core. This, in effect, makes the Earth a giant electromagnet. The magnetic field created by the internal motions of the core produces the magnetosphere. This protects the Earth's atmosphere from the solar wind. As the Earth is 4.6 billion years old, it would have lost its atmosphere by now if there were no protective magnetosphere.

Above the crust is the Earth's atmosphere. Water vapours and carbon dioxide allow the Earth's atmosphere to catch and hold the Sun's energy through a phenomenon called the greenhouse effect. This allows Earth's surface to be warm enough to have liquid water and support life.

ACTIVE EARTH

Enveloping Earth like a giant jigsaw, tectonic plates have gradually created the familiar features of our planet's surface. Plates spread and collide, forming mountains, volcanoes and valleys in an ongoing process. Moving at just a few centimetres a year plate tectonics may seem slow but over millions of years they have moved thousands of kilometres, reshaping continental landmasses into the form we see on our maps today. Earth's highest mountains, the Himalaya, are on the boundary of the Eurasian Plate and Indo-Australian Plate – they were created by the collision of the plates 65 million years ago and are still rising in height by about 6 cm a year. The distribution of earthquakes and volcanoes is closely related to plate boundaries. Today there are hundreds of active volcanoes on our planet, and probably more unknown ones under the ocean's surface, but during the lifespan of the Earth so far it is likely that there have been many millions more.

The structure of Earth

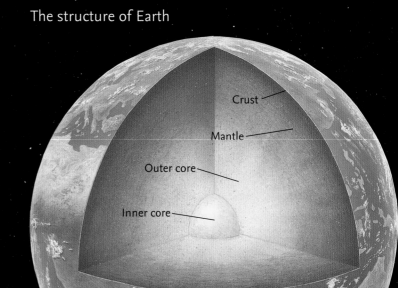

Crust

Mantle

Outer core

Inner core

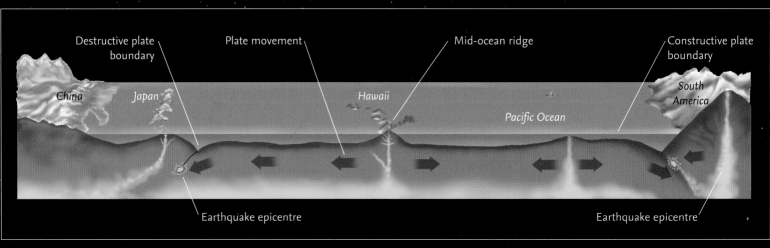

Destructive plate boundary

Plate movement

Mid-ocean ridge

Constructive plate boundary

China

Japan

Hawaii

South America

Pacific Ocean

Earthquake epicentre

Earthquake epicentre

A cross-section through the Pacific Ocean showing the plate structure

Earth's major plates

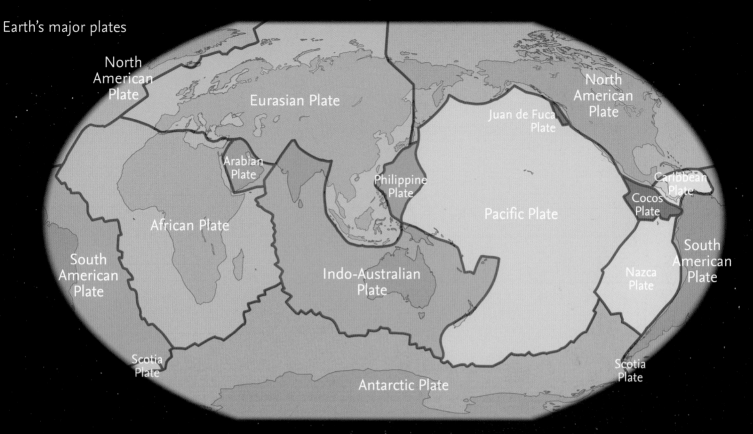

North American Plate

Eurasian Plate

North American Plate

Juan de Fuca Plate

Arabian Plate

Philippine Plate

Caribbean Plate

African Plate

Cocos Plate

Pacific Plate

South American Plate

South American Plate

Indo-Australian Plate

Nazca Plate

Scotia Plate

Scotia Plate

Antarctic Plate

Loss of life in the ten most deadly earthquakes since 1900

1976	China	255 000
2004	Indonesia/Indian Ocean	225 000
2010	Haiti	222 570
1920	China	200 000
1927	China	200 000
1923	Japan	142 807
1908	Italy	110 000
2005	Pakistan	74 648
1932	China	70 000
1970	Peru	66 794

Mt Etna, on the island of Sicily in Italy, lies on the boundary between the Eurasian Plate and the African Plate. It is one of the most active volcanoes in the world

EARTH — FACTS

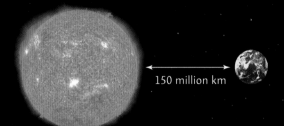

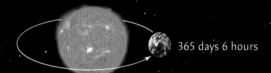

150 million km

Distance from Sun

365 days 6 hours

Length of year

23 hours 56 minutes

Length of day

12 700 km

Diameter

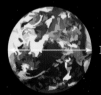

40 000 km

Circumference

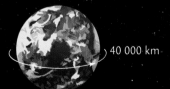

15 °C

Average temperature

1

Number of moons

Earth's speed

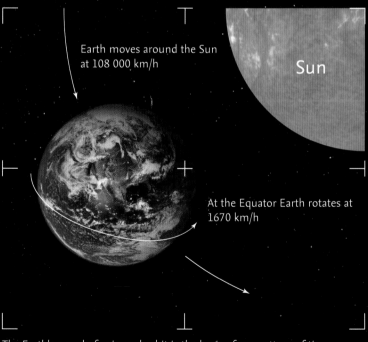

Earth moves around the Sun at 108 000 km/h

Sun

At the Equator Earth rotates at 1670 km/h

The Earth's speed of spin and orbit is the basis of our pattern of time. Our calendar year is based on the time it takes for Earth to complete its 1 billion kilometer orbit of the Sun. The 24 hour length of our day is based on the time it takes for Earth to spin once on its axis

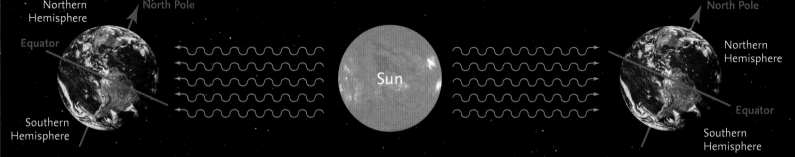

Summer is warmer than winter (in each hemisphere) because the Sun's rays hit the Earth at a more direct angle during summer than during winter and because the days are much longer than the nights during the summer. During the winter, the Sun's rays hit the Earth at an extreme angle, and the days are very short.

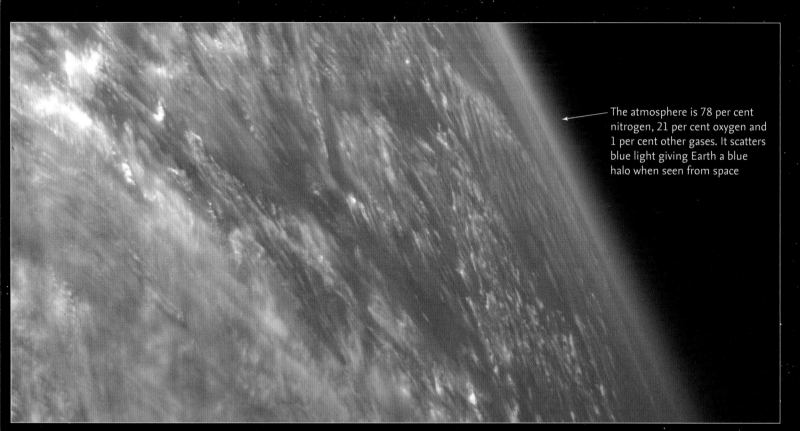

The atmosphere is 78 per cent nitrogen, 21 per cent oxygen and 1 per cent other gases. It scatters blue light giving Earth a blue halo when seen from space

Earth's atmosphere is a layer of gases about 450 km thick that surrounds the planet. All our weather is formed in the lowest 10 km of this layer

EARTH'S MOON

The Moon is Earth's only natural satellite. It is the second brightest object in the sky after the Sun. It is the only object, other than Earth, to have been stepped on by human beings.

You can clearly see, particularly with binoculars, many dark patches on the Moon's surface. Ancient astronomers thought these were filled with water, so they are known as *mare*, after the Latin for 'sea'. We now know they are huge solidified pools of lava. There is no atmosphere and no oceans on the Moon, but ice left by crashed comets has been found.

Other parts of the Moon's surface are very mountainous and there are peaks nearly as high as Mount Everest (the highest mountain on Earth). The Moon is dotted with craters made by meteorite impacts that mostly happened 4 billion years ago. Earth was similarly bombarded at this time, but its atmosphere, rain, wind and water have eroded most of them.

Imagine a planet the size of Mars crashing into the Earth. This is what happened 4.5 billion years ago. The cosmic mash-up generated huge amounts of heat: much of both planets fused together and a huge lump of molten debris was flung out into space. This lump of debris became the Moon. Gravity eventually helped both the Earth and the new Moon pull themselves into spheres.

The Moon changes its appearance to us in a regular monthly pattern, or phases. These phases happen because the Moon orbits the Earth once every 27.3 days. When the Moon passes between the Sun and Earth we can see very little of the sunlight that hits it, causing a new moon phase. Fourteen days later the Moon is on the far side of the Earth from the Sun and we see it completely illuminated. This is a full moon. The lunar phase cycle at 29.5 days is longer than the orbital period of the Moon, this is because during the Moon's orbit around the Earth the Earth is also moving around the Sun and we have to wait a little longer to see the same phase in the sky.

Although the Moon's appearance changes, we always see the same side of the Moon. This is because it turns on its axis in exactly the same time it takes to go around the Earth. This is called synchronous rotation. The other side is not 'dark', however. Also, although the Moon seems bright to us, its surface is really about as reflective as a lump of coal. Its perceived brightness is due to contrast with the surrounding dark space.

The Moon is 3,476 km in diameter, a quarter the diameter of the Earth. This makes it the largest satellite in the Solar System relative to its planet. Its mass gives it a gravitational pull strong enough to affect our oceans. Tides are highest when at full and new moons. The Moon appears to be same size in the sky as the Sun, allowing for total solar eclipses. This is a coincidence.

Between 1969 and 1972 six Apollo missions landed on the Moon, with twelve astronauts stepping on its surface. You can see from pictures of their missions that the sky on the Moon is always black, even during the day. This is because it has no atmosphere to scatter sunlight. The astronauts wore heavy spacesuits to provide air to breathe and to protect them from the sun's more intense radiation. But since the Moon's gravity is one sixth of that on Earth they were still able to leap like gazelles!

THE MOON — EARTH'S COMPANION

The Moon has been the Earth's companion and the largest presence in our night sky for several billion years. Its Roman name is Luna and its Greek name is Selene, and its rhythm has been part of our timekeeping for thousands of years. Its synchronous rotation means that although it has been observed for so long we knew nothing about its far side until a probe first photographed it in 1959. Manned Apollo orbits in the late 1960s and early 1970s increased our knowledge of the far side, revealing a densely cratered hemisphere with more highlands and fewer of the darker 'seas' than the near side. This indicates a thicker crust where lava could not so easily rise to the surface.

A blue moon is when there are two full moons in one month, one at the very beginning of the month and one at the end. The last blue moon was 31 August 2012 and the next is 31 July 2015.

The Moon's orbital plane is inclined by about 5° in relation to the Earth's

Moon's orbit

Earth's orbit

A lunar eclipse occurs when the Earth passes between the Sun and Moon, and the Earth's shadow falls on the Moon. A total eclipse makes the Moon appear a reddish colour, as in this image. The next total eclipse of the Moon is in April 2014

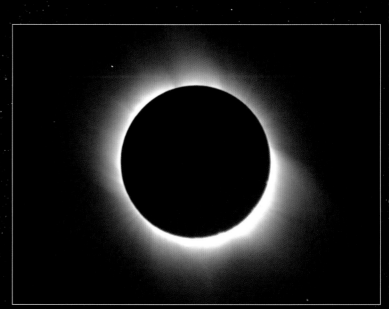

A solar eclipse occurs when the Moon passes between the Sun and Earth, and the Moon's shadow falls on Earth. In this image of a total solar eclipse only the halo of the Sun's corona is visible around the edge of the Moon

The phases of the Moon

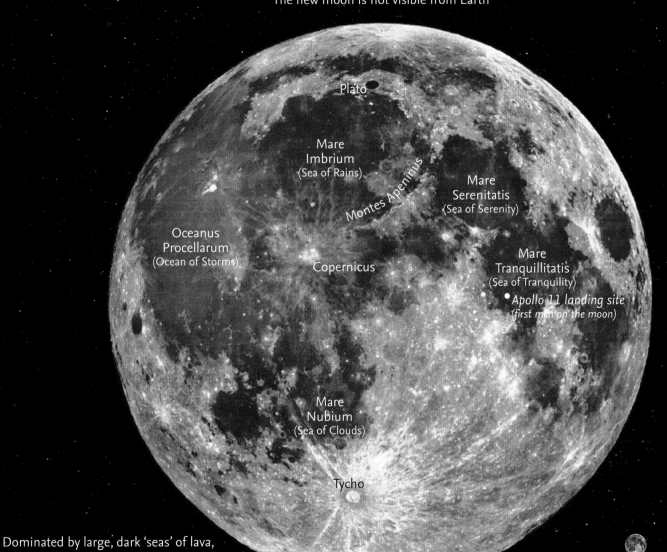

New moon   Waxing crescent   First quarter   Waxing gibbous   Full moon   Waning gibbous   Last quarter   Waning crescent   New moon

The new moon is not visible from Earth

Plato

Mare Imbrium (Sea of Rains)

Montes Apeninus

Mare Serenitatis (Sea of Serenity)

Oceanus Procellarum (Ocean of Storms)

Copernicus

Mare Tranquillitatis (Sea of Tranquility)

• Apollo 11 landing site (first men on the moon)

Mare Nubium (Sea of Clouds)

Tycho

Dominated by large, dark 'seas' of lava, and the prominent ray crater of Tycho in

EARTH'S MOON — FACTS

Distance from Earth

384 000 km

27 Earth days
7 hours 43 minutes

Orbital rotation period

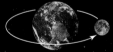

29 Earth days
12 hours 44 minutes

Synodic rotation period

3476 km

Diameter

10 921 km

Circumference

-173°C to 123°C

Average temperature

Jim Irwin driving the 1971 Apollo 15 lunar rover already prepared for the first lunar spacewalk

The far side of the Moon was first photographed in 1959, by the
Soviet Union probe called Luna 3. This image mosaic was created
using images taken by the Lunar Reconnaissance Orbiter
Wide Angle Camera between 2009 and 2011.
This side of the Moon has many craters
as it is always exposed to impacts,
while the near side is shielded
by the Earth.

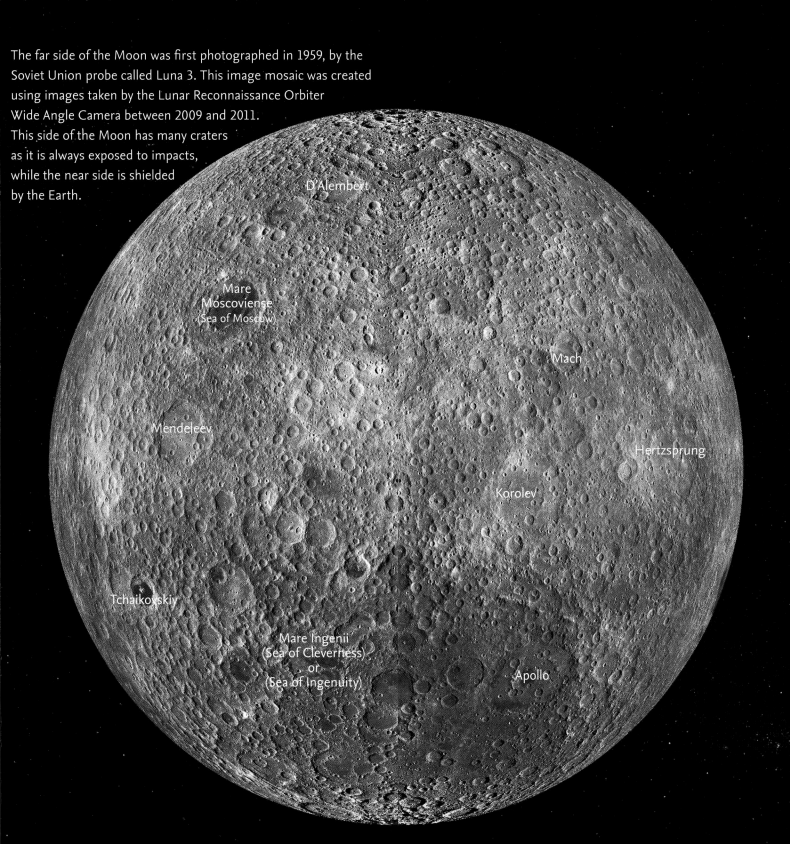

D'Alembert

Mare
Moscoviense
(Sea of Moscow)

Mach

Mendeleev

Hertzsprung

Korolev

Tchaikovskiy

Mare Ingenii
(Sea of Cleverness)
or
(Sea of Ingenuity)

Apollo

MARS

Mars is the fourth planet from the Sun and the second smallest in the Solar System. It looks blood red and so was named after the Roman god of war. This colour is really caused by the oxidised iron in its surface soil – Mars is literally rusting!

Like Earth, Mars has an atmosphere, but it is thin and mostly of carbon dioxide. Some of this has condensed to form polar ice caps. These vary in size depending on the season. Mars has the largest dust storms in the Solar System. A gigantic dust storm obscured the entire surface of the planet for three whole months in 2001.

A Martian day is almost the same length as Earth's and it is tilted on its axis at a similar angle to Earth. This means that it experiences seasons like we do – although they are much more extreme. Martian surface temperatures vary from an icy winter lows of about –143 °C to summer highs of 35 °C. Martian seasons are twice as long as ours, as Mars's greater distance makes the Martian year about two Earth years long.

There is a giant crack in the skin of the planet – the Valles Marineris. This snakes a third of the way round Mars and at 4,000 km long would stretch across the USA. At 7 km deep, you could lose all but the highest mountains on Earth in it.

Mars has many other awesome geographical features. Olympus Mons is the largest of several immense volcanoes. It reaches 26 km up into the Martian atmosphere, as high as three Mount Everests stacked on top of each other. At its rim is a sheer cliff 8 km high. It is the tallest mountain on any planet in the Solar System.

Mars has two moons, Phobos and Deimos. They are potato-shaped rather than round. Phobos is slowly spiralling towards the Martian surface. In 40 million years it will finally hit the planet. If any of our distant descendants are working or even living on Mars, they will be able to watch a moon fall from the sky. Probably not a good idea!

The gravity of Deimos is 2,000 times less than on Earth. If you were a space-suited astronaut standing on Deimos, you could easily jump a full kilometre vertically, before drifting slowly back to the surface.

Scientists have found dried-up rivers on Mars, although water has long ceased to flow. The atmosphere was once thicker, and supported wet and warmer conditions, but the planet's magnetic field disappeared 4 billion years ago, probably because its molten core cooled. The atmosphere was exposed to the brutal solar wind and the oceans boiled off into space.

However, where there was once water, there could have been life. Perhaps there is still microbial life within rocks or under the planet's surface. NASA's car-sized Curiosity rover began exploring the planet in 2012, trying to find out if the planet could ever have supported life. It has a laser that can zap rocks up to 7m away into dust so it can analyse the particles. Also on board is a drill that can dig into rocks, a microscope and several cameras. Curiosity is sure to make many amazing discoveries in future.

MARS — THE RED PLANET

Once thought to be the most likely planet upon which we would find other life forms, it was a surprise when the first flyby of Mars in 1965 revealed a dramatic landscape but one, nevertheless, devoid of obvious signs of life. Most of the planet is covered by red-brown rocks and layers of dust but it is not a uniform surface. Over time it has been affected by volcanoes, crustal movements and giant dust storms. Darker parts of the planet, once thought to be areas of vegetation by observers from Earth, are areas of darker rock lacking surface dust and these change in shape each year as dust is blown about. The south of the planet is generally less low-lying than the north, with rocks that are generally older and a land surface that is more cratered. Thin clouds of water ice hang over the planet, especially over higher land, and there is water in the polar ice caps but no running water to be found.

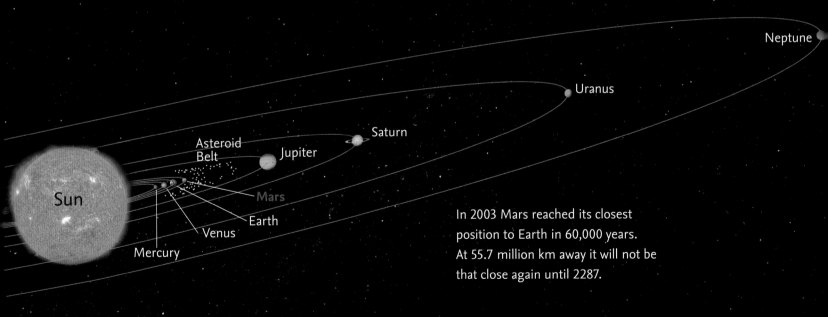

Sun

Mercury

Venus

Earth

Mars

Asteroid Belt

Jupiter

Saturn

Uranus

Neptune

In 2003 Mars reached its closest position to Earth in 60,000 years. At 55.7 million km away it will not be that close again until 2287.

Mars' polar ice caps expand and contract with the seasons, growing in summer and melting in winter. They comprise a mixture of water ice and carbon dioxide ice. Cutting into the north polar ice cap is the Chasma Boreale, a 560 km-long canyon

Low-lying and gently-sloping Alba Mons is Mars's widest spreading volcano with lava fields that stretch for over 1,350 km

The second highest known mountain in the Solar System, and the highest volcanic mountain, the extinct shield volcano of Olympus Mons has a crater over 80 km wide. It last erupted about 25 million years ago

The ridge of Tharsis has three large volcanoes - from north to south they are Ascraeus, Pavonis and Arsia

The Valles Marineris, just south of the equator, is the largest of the planet's canyons and gullies and was probably formed by a cracking in the planet's crust similar to the Rift Valley in East Africa. Partly obscured by cloud in this image is an ancient riverbed winding northwards from the main canyon system

MARS — FACTS

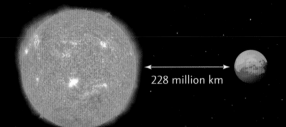

Distance from Sun

228 million km

Average temperature

-63 °C

Length of year

687 Earth days

Number of moons

2

Length of day

24 hours 37 minutes

Diameter

6779 km

Circumference

21 300 km

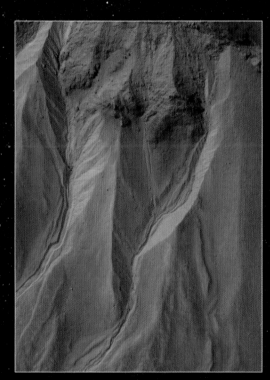

These gullies, V-shaped channels running downhill, are near Hale crater in the southern hemisphere. Scientists hope that if they monitor the gullies they will find out if there is any liquid water on Mars

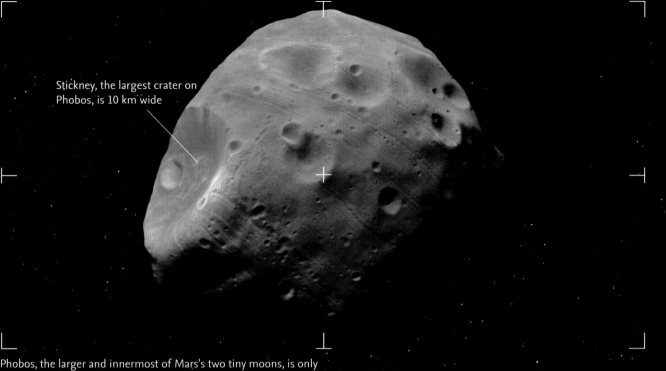

Stickney, the largest crater on Phobos, is 10 km wide

Phobos, the larger and innermost of Mars's two tiny moons, is only 27 km across. Both Phobos and Deimos, Mars's other moon, are thought to be asteroids that were captured into orbit. Their masses are too low for gravity to make them spherical

An image of Mars's surface taken by the Mars Pathfinder Lander in 1997 in the Ares Vallis, a valley region believed once to have flowed with liquid that may have been water.

ASTEROIDS

Asteroids are lumps of rock and ice that orbit the Sun like mini-planets. They vary in size from dust specks to lumps the size of countries. There are millions of asteroids and most of them are concentrated in a zone just outside the orbit of Mars, the asteroid belt. Sir William Herschel coined their name in 1802 because they appeared 'star-like'.

Around 500,000 asteroids have been identified, with thousands more being added to the catalogue every year. The asteroid Ida is 52 km long and has its own tiny moon, Dactyl. Vesta is the brightest asteroid visible from Earth and is 525 km in diameter. It boasts the largest mountain in the Solar System, an impact crater that stretches 31 km from floor to the top of its rim.

Some asteroids orbit further out than the main belt. The Trojans mostly orbit at the same distance as Jupiter although some have been found to share the orbits of Earth, Mars and Neptune, while the Centaurs are even further away, and are like a cross between comets and asteroids.

Like most other small bodies, asteroids are the shattered remnants of objects within the young Solar System. It's likely that the asteroid belt formed as a group of planetesimals, the smaller building blocks of the planets. These may have got as far as forming a small planet that was later smashed to pieces by a planetesimal.

Alternatively, Jupiter's powerful gravitational pull may have acted like an enormous vacuum cleaner, yanking these baby planets apart and smashing them into each other, preventing them from ever forming a planet.

Many were flung out of the area altogether, and today the total mass of all asteroids is only one twentieth that of the Moon. But in the first 100 million years of the Solar System there may have been 1,000 times as many.

Like a pair of cosmic bullies, Mars and Jupiter still fling their weight around the asteroid belt, with their gravitational pull causing asteroids to collide. Slow collisions can join two asteroids together. Fast collisions can break them into 'families' of smaller fragments. Sometimes these fragments hurtle into the inner Solar System where they can hit planets, including Earth.

Millions of meteorites land on our planet every year without incident. But if a meteorite is big enough and moving fast enough it will form a crater. If it is really big it can cause tremendous damage. Around 65 million years ago an asteroid or comet around 10 km across hit the Earth near what is now Mexico, creating the Chicxulub crater, 180 km wide. This massive impact created global firestorms, huge tsunamis, plunged the Earth into darkness for a year and killed 80 per cent of all species, including the dinosaurs.

Astronomers constantly monitor the skies for asteroids big enough to threaten our planet. They have found at least 1,000 Near-Earth Asteroids more than 1 km in diameter. Although these would do considerable damage if they hit us, an impact with such a big asteroid is only likely to happen once every 100,000 years.

ASTEROIDS

Amongst the 500,000 asteroids that have been identified there are an amazing variety of forms. Larger asteroids are mainly spherical whilst smaller ones can have very irregular shapes. The largest, 4 Vesta, is the only one that is ever visible to the naked eye. The Rheasilvia crater on 4 Vesta covers most of its southern hemisphere – in 2011 the central peak of the crater was discovered to be the highest mountain in the Solar System.

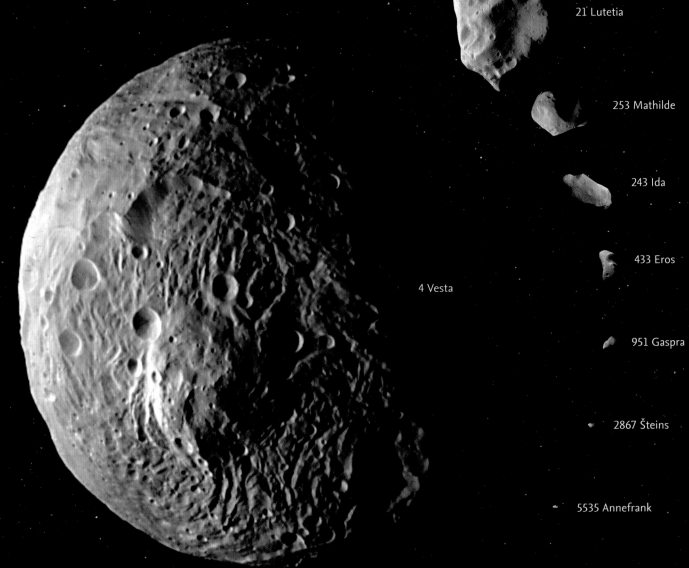

21 Lutetia

253 Mathilde

243 Ida

433 Eros

4 Vesta

951 Gaspra

2867 Šteins

5535 Annefrank

25143 Itokawa

Asteroids come in many shapes and sizes

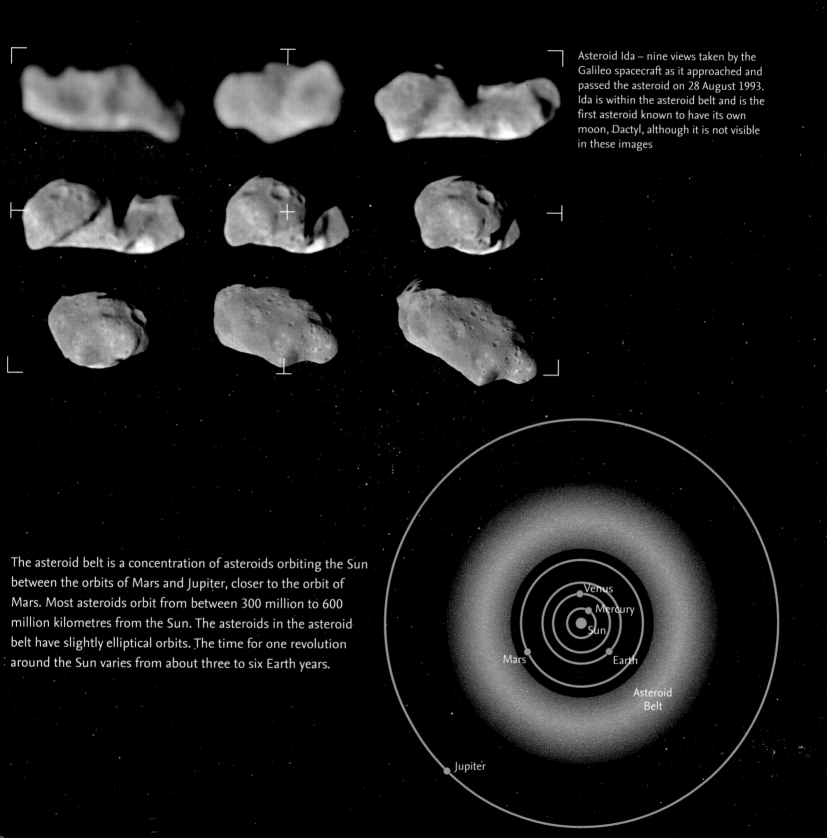

Asteroid Ida – nine views taken by the Galileo spacecraft as it approached and passed the asteroid on 28 August 1993. Ida is within the asteroid belt and is the first asteroid known to have its own moon, Dactyl, although it is not visible in these images

The asteroid belt is a concentration of asteroids orbiting the Sun between the orbits of Mars and Jupiter, closer to the orbit of Mars. Most asteroids orbit from between 300 million to 600 million kilometres from the Sun. The asteroids in the asteroid belt have slightly elliptical orbits. The time for one revolution around the Sun varies from about three to six Earth years.

THE OUTER SOLAR SYSTEM

The outer planets are Jupiter, Saturn, Uranus and Neptune. They are sometimes known as the Gas Giants as they are huge in comparison with the inner planets and, made up mostly of gas, do not have solid surfaces. All four have rings around them, with Saturn's being the most famous. The four planets also have large numbers of moons orbiting them.

The area beyond Neptune, called the 'trans-Neptunian region', is largely unexplored. Most of its residents are small worlds of rock and ice far less massive than our Moon.

The Kuiper belt is this area's first gathering of such objects. It is a great ring of debris similar to the asteroid belt, made mostly of tiny icy lumps but also containing dozens of dwarf planets. The dwarf planet Pluto is the largest known object in the Kuiper belt.

When Pluto was discovered in 1930 it was considered the ninth planet, but it was demoted to dwarf planet status in 2006.

Makemake is the brightest object in the Kuiper belt after Pluto. It was named and designated a dwarf planet in 2008. Haumea is about the same size as Makemake and has two natural satellites.

The scattered disc overlaps the Kuiper belt but extends much further outwards. This is the breeding ground for many comets. Eris is the largest known scattered disc object and the most massive of the known dwarf planets. It has made astronomers debate what a planet is, since it is 25 per cent more massive than Pluto and about the same diameter. It has one known moon, Dysnomia.

Where does the Solar System end? Its outer boundary is not precisely defined. The solar wind's influence continues out to four times Pluto's distance from the Sun. This is the edge of the heliosphere, the bubble in space that surrounds the Sun. It is considered the beginning of the interstellar medium, the area of space between galaxies. But the Sun's gravity is felt much further out.

The Oort cloud is a spherical shell of up to a trillion icy objects that has been proposed but not observed. It is thought to surround the Solar System from a distance of around 1 light-year and contains comets that were ejected from the inner Solar System by the gravitational pull of the outer planets.

THE OUTER SOLAR SYSTEM

The cold, dark outer reaches of the Solar System are the least explored. Without a solid crust it is not possible to land on any of the outer planets. From the orbits of these Gas Giants our home planet appears the same size as a star in the sky appears to us here on Earth. At the far outer limit of the Solar System, around 14 billion km away, the Earth would be invisible with the naked eye and the Sun itself would appear merely as the brightest star in the sky. Launched in 1977, Voyager 2 is designed to explore the outer Solar System and is the only spacecraft so far to study all four of the outer planets. It is on a course taking it out of the Solar System, hopefully returning data until at least 2025.

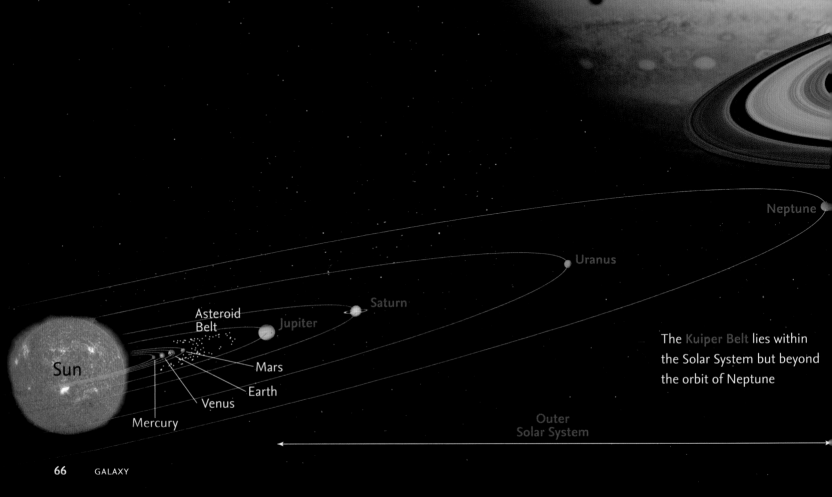

Neptune

Uranus

Saturn

Asteroid
Belt

Jupiter

Sun

Mars

Earth

Venus

Mercury

The Kuiper Belt lies within the Solar System but beyond the orbit of Neptune

Outer
Solar System

Jupiter
The fifth planet from the Sun and the largest planet, with a mass more than twice the size of all the other planets combined.

Saturn
The second-largest but least dense of all the planets, Saturn is surrounded by spectacular rings of ice and rock which may be less than 10 m thick in places.

Uranus
The third-largest planet in the Solar System, and the first planet to be discovered with a telescope.

Neptune
Named for the Roman god of the sea, Neptune is the fourth-largest planet and the most distant one from the Sun.

JUPITER

Jupiter is the king of the planets. It has a mass 2.5 times greater than that of all the other planets combined and a diameter of 142,984 km at its equator – equivalent to eleven Earths. Fittingly, it is named after Jupiter, chief Roman deity and god of the sky and thunder. The planet is the fourth brightest object visible from Earth after the Sun, the Earth's Moon and Venus.

We only see Jupiter's outer atmosphere from Earth. This is a swirling, poisonous mix of methane and ammonia. It is clearly divided into striped bands of clouds. Where the bands meet, huge storms rage. Violent winds and lightning a thousand times as powerful as lightning on Earth pummel the atmosphere.

Like on Earth, Jupiter's storms are fed by heat. When our storms move from the sea to land they lose this heat

energy. But there is no land on Jupiter so storms power on and on. The Great Red Spot is a giant storm resembling an eye that has been blowing for at least 300 years. You could fit the entire Earth into this storm twice over.

You couldn't land a spacecraft on Jupiter. When you dropped into the clouds you would soon be crushed by the pressure of the planet's immense gravity. Even if you did make it through the outer atmosphere you would find that the planet has no solid surface like the Earth and other terrestrial planets. Beneath your feet would be an ocean of liquid hydrogen thousands of kilometres deep.

Although classed as a gas planet, in many ways Jupiter is like a tiny star that didn't become quite big enough to shine. Like a star it is mostly composed of hydrogen and helium. It probably has a rocky core of heavier elements surrounded by a deep layer of liquid metallic hydrogen, crushed to incredible density by the planet's gravity. Outside this is a thinner layer of liquid and gaseous hydrogen.

Jupiter spins the fastest of all planets, completing a rotation on its axis in less than ten hours. Like the other gas planets, it has a ring system. Unlike Saturn's wide rings of bright ice, Jupiter's are faint and made of tiny dust particles. Orbiting Jupiter are at least sixty-seven moons, including the four large moons: Io, Europa, Ganymede and Callisto discovered by Galileo in 1610. Ganymede, the largest of these moons, is bigger than the planet Mercury.

Jupiter's huge mass and position near the inner solar system make it a regular target for comets. In 1994 the Galileo orbiter spacecraft took spectacular pictures of the fragments of comet Shoemaker-Levy 9 smashing into Jupiter.

JUPITER — THE GIANT PLANET

Jupiter never comes closer to Earth than about 600 million kilometres away but its immense size means that it is still the fourth brightest object in the night sky. Yet it is only Jupiter's outer atmosphere that is visible from Earth. Most of Jupiter is comprised of hydrogen and helium although the tops of its clouds are made mainly of ammonia. Other chemicals, including phosphorous, sulphur and hydrocarbons, also rise to the surface creating a colourful composition of yellow, brown, red and white. Jupiter spins so fast that it bulges out at its centre and these clouds are drawn out into colourful bands. Winds of up to 450 km/h also blow in the upper atmosphere creating beautiful and constantly changing swirling patterns.

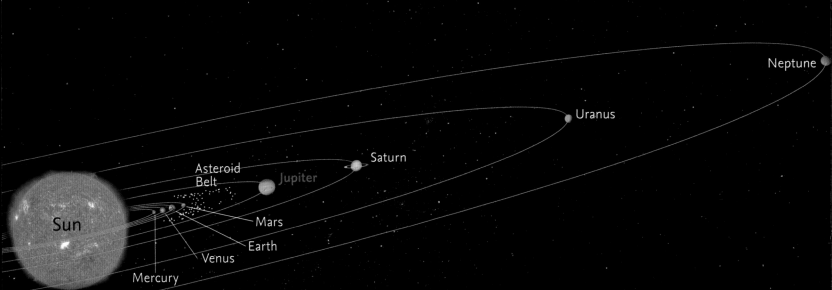

Sun

Mercury

Venus

Earth

Mars

Asteroid Belt

Jupiter

Saturn

Uranus

Neptune

Bluish areas are those
with the lowest cloud cover

Small bright clouds show
areas of violent lightning storms

Shadow of Europa, one of
Jupiter's moons, on the cloud
tops of the planet's atmosphere

The Great Red Spot is a giant
atmospheric storm that has
raged for over 300 years.
It is trailed to the west by a
turbulent region

A true colour image of
Jupiter taken by Cassini in 2000.

Earth in comparison

JUPITER'S MOONS

The Sun is a star with eight planets orbiting it. Jupiter also seems to have its very own solar system with at least sixty-seven very different moons orbiting it. Most of these moons are very small and are probably asteroids caught by Jupiter's strong gravitational pull keeping them in orbit around the planet.

In 1610, Italian scientist Galileo Galilei turned one of the earliest telescopes towards Jupiter and almost immediately became the first human to see the planet's four largest moons. These were the first objects found to orbit a body that was neither Earth nor the Sun. His findings threatened the Roman Church's belief that the Earth was at the centre of the Universe. Galileo was tried and put under house arrest for the rest of his life.

The four largest moons, Io, Europa, Ganymede and Callisto are known as the Galilean satellites and you can see them with binoculars on a clear night. They are like little worlds, each different from the other.

Io is the closest moon to Jupiter and is pulled by huge gravitational forces as it orbits. Io's sister moons Ganymede and Europa also tug at Io. The result? Poor Io is constantly squeezed and stretched by tidal forces, like a squash ball in your fist. Its surface can bulge by over 50 m a day. The friction inside the moon fuels spectacular volcanic plumes 500 km high. The moon looks like a pizza, and its topping changes almost daily! Io also generates a huge electrical charge that sparks tremendous storms.

Contrastingly, icy Europa is one of the smoothest bodies in the Solar System. It's likely that beneath its 30 km thick crust of solid ice is liquid ocean up to 100 km deep. This ocean is warmed by internal heat similar to that driving Io's volcanoes and holds twice the volume of water in all Earth's oceans. It is a prime candidate for extra-terrestrial life – microbes could exist here as they do at hydrothermal vents in Earth's deep oceans.

Ganymede is the largest moon in the Solar System and is bigger than the planet Mercury. It is the only moon with its own magnetic field. Ganymede has a thin oxygen atmosphere and a saltwater ocean trapped 200 km beneath its surface. Callisto's surface is extremely old and is one of the most heavily cratered in the Solar System.

The Galilean satellites and four other small, close-in moons are all regular satellites with nearly circular orbits. The others are much further out and have more eccentric orbits. They are called irregular satellites.

Some moons whizz round Jupiter in as little as seven hours (taking less time than Jupiter does to spin around its axis). Others take almost three Earth years to orbit the planet.

JUPITER'S MOONS

When Galileo Galilei discovered Jupiter's four largest moons they were the first moons discovered in the Solar System other than the one that orbits Earth. Our moon was known as 'the Moon' but these new moons needed names of their own and were called after the lovers of the mythological character Zeus (Jupiter is the Roman name for Zeus). They are visible with ordinary modern binoculars only as tiny points of light. The next of Jupiter's moons to be discovered was not until 1892 when red-coloured Amalthea became the last moon to be discovered by direct observation. Subsequent discoveries of Jupiter's moons in the twentieth and twenty-first century have all depended on telescopic photography and space probes. It is possible that there are still more tiny moons to be found.

Jupiter and the four moons discovered in 1610 by Galileo Galilei

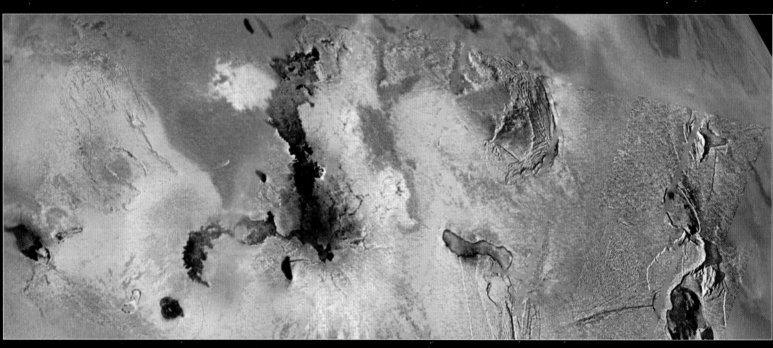

An image of the mountains and volcanoes on Io taken by the Galileo spacecraft in 1999. The mountain near bottom-right is about 8 km high and the bright white areas near the centre of the image are believed to be sulphur-dioxide

Io, with over 400 active volcanoes, is the most geologically active object in the Solar System. Probes have photographed volcanic plumes of gas and dust rising up to 500 km high.

Ganymede is the largest moon in the Solar System.and is larger than the planet Mercury.

Europa is the smallest of the Galilean moons and orbits Jupiter in just over three and a half days.

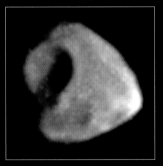

Thebe, one of Jupiter's smallest moons, is about 100 km across and takes just over 16 hours to orbit Jupiter

Callisto is the most heavily cratered object in the Solar System and with a surface that has hardly changed for over 4 billion years it can also claim to have one of the oldest landscapes in the Solar System.

JUPITER — FACTS

778 million km

Distance from Sun

-148 °C

Average temperature

11 Earth years 314 days

Length of year

67

Number of moons

9 hours 56 minutes

Length of day

Liquid and
gaseous hydrogen

140 000 km

Diameter

The rocky core

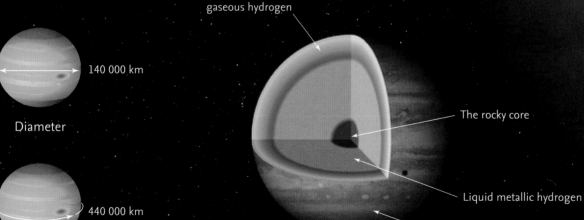

440 000 km

Circumference

Liquid metallic hydrogen

Atmosphere

Interior of Jupiter

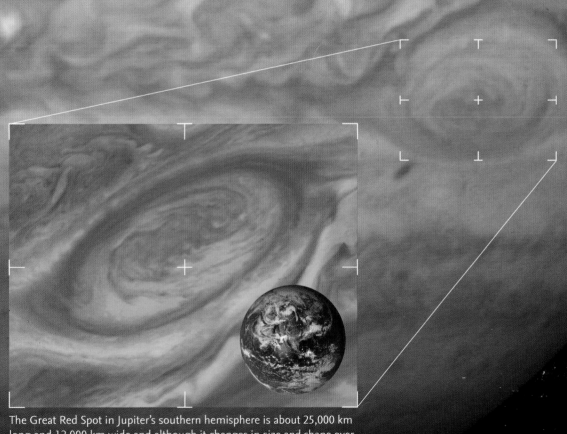

The Great Red Spot in Jupiter's southern hemisphere is about 25,000 km
long and 12,000 km wide and although it changes in size and shape over
time it is still the only one permanent feature of the planet

SATURN

Saturn is the second largest planet in the Solar System after Jupiter. It is named after the Roman god of agriculture and is also known as the ringed planet because it is encircled by majestic rings of water ice particles. Saturn is a gas giant with an average diameter about nine times that of Earth.

Like Jupiter, Saturn is made up of mostly hydrogen and helium, and has no solid surface that you could stand on. The planet that we see is really just its cloudy upper atmosphere, which appears pale yellow due to ammonia crystals. As Saturn spins, these clouds are whipped into horizontal stripes encircling the planet. Wind speeds on Saturn can reach a fearsome 1,800 km/h. Huge storms often appear in the atmosphere. A long-lasting storm at the planet's north pole is hexagonal in shape and is several times larger than the Earth.

Saturn has a very hot interior, reaching 11,700 °C at its rocky core. This is surrounded by a deep layer of metallic hydrogen, an intermediate layer of liquid hydrogen and liquid helium and an outer gaseous layer.

Saturn spins very quickly. Its day lasts for just ten hours, the time it takes to rotate fully. But its year is long: Saturn takes almost thirty Earth years to go round the Sun. If you could find a pond large enough to drop Saturn into, you would find the planet floated! This is because, although it has a solid core, it is mostly gaseous and is 30 per cent less dense than water.

Every astronomer remembers the first time they saw Saturn through a telescope. Its swooping, dazzling rings are one of the most beautiful sights in the night sky. Saturn's rings aren't solid bands but countless millions of water ice shards, with tiny amounts of rocky material. These shards vary in size from microscopic particles to pebbles up to boulders the size of your house. The ring system either formed at the same time as Saturn, or later when some of its moons smashed into each other and fractured into fragments. There are still hundreds of moonlets within the rings.

The ring system is at least 270,000 km in diameter, two-thirds of the distance from Earth to the Moon. But the rings are incredibly thin, averaging just 20 metres deep – the height of a six-storey building. The rings themselves are much more complicated than they first appear. Gravity twists them into swirls and creates distinct gaps between them. The lumps of ice in the rings are constantly clumping together and then being blasted apart by collisions. This keeps the ice surfaces clear of dust, so the rings reflect sunlight and appear bright to us on Earth.

SATURN — THE RINGED PLANET

Although it is a whirling ball of gas Saturn usually presents a rather dull, cloudy face into space. There is a banding pattern caused by high winds but no other prominent features. However, about every thirty years there are large bright patches of clouds in the atmosphere of its northern hemisphere. Called Great White Spots, they usually stretch to a width of several thousand kilometres and are large enough to be seen by a telescope on Earth. In 1990 an area that started as a white spot extended into storms that spread to encircle the whole planet. They are spring and summer seasonal features, made of ammonia particles that are pushed up through the tops of the clouds by warm air, and they occur so infrequently because of the length of time that it takes for Saturn to orbit the Sun.

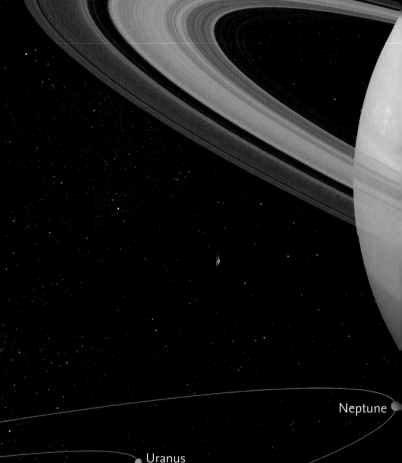

A huge storm, called a Great White Spot, started in late 2010, by the end of February 2011 it had travelled right around the planet

Neptune

Uranus

Saturn

Asteroid
Belt

Jupiter

Sun

Mars

Earth

Venus

Mercury

Clouds over Saturn's surface are
masked by a layer of yellowish haze

The space between Saturn's surface
and the inner ring is three times the
diameter of the Earth

Saturn's rings are one of the most spectacular
and beautiful sights in the Solar System.

Earth in comparison

SATURN'S RINGS

Saturn's rings are probably the most distinguishing feature of any planet in the Solar System. All four Gas Giants have rings but only Saturn's are so spectacular. The width of the planet is 116,500 km, but the rings surrounding it increase this width to around 270,000 km. Although first seen by Galileo through a telescope and then identified as rings later in the seventeenth century, it was not until more recently that their complexity was revealed. In the 1980s Voyager 1 and Voyager 2 discovered that they are composed mainly of water ice and small amounts of rocky material. Particle size ranges from tiny dust-sized grains to large lumps and occasionally mountain-sized pieces. It is the reflection of sunlight that makes the rings so visible. Blocking light from the Sun, they also cast shadows down onto the planet below.

The changing tilt of Saturn as it orbits the Sun makes the rings appear to open and close over time when viewed from Earth, as this sequence of images taken from 1996–2000 shows

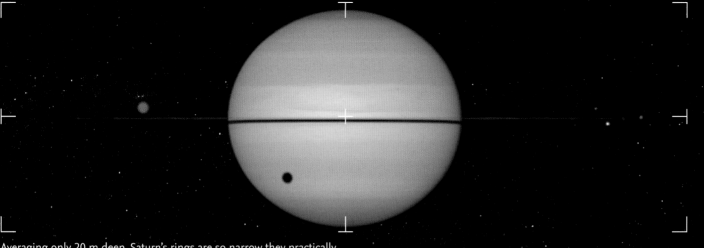

Averaging only 20 m deep, Saturn's rings are so narrow they practically
disappear when viewed 'edge' on

Saturn's rings are divided into eight major ring divisions and each ring is
made up of thousands of narrow rings, called ringlets. Each ring orbits at a
different speed around the planet

SATURN'S MOONS

There are at least sixty-two moons orbiting Saturn. These range from tiny moonlets less than one kilometre across to the enormous Titan, which is larger than the planet Mercury.

In terms of size, Titan is the big daddy of Saturn's moons, making up 96 per cent of the mass in orbit around the planet. All the other moons combined make up the remaining 4 per cent.

Titan is the only moon in the Solar System to have a thick atmosphere. It spends 95 per cent of its time within Saturn's magnetosphere, helping to shield its atmosphere from the solar wind. The atmosphere is rich in nitrogen, like Earth's atmosphere, with some methane and hydrocarbons. The hydrocarbons may be created when methane is broken down by the Sun's ultraviolet light, producing a dense orange smog. Many of these molecules were the building blocks of life on Earth and Titan is a hot contender for hosting life beyond our planet.

Titan has a climate with wind, rain and seasonal weather patterns that create surface features similar to those of Earth. These include dunes, rivers, lakes and seas. But Titan's weather is based on methane, not water. Titan's methane cycle mirrors Earth's water cycle but at a much lower temperature.

In 2004, the Cassini–Huygens space probe entered into orbit around Saturn. Cassini captured radar images of large, liquid hydrocarbon lakes and their coastlines with numerous islands and mountains. The lakes were the first stable bodies of surface liquid found outside of Earth. The orbiter then released the Huygens probe that landed on the surface of Titan in 2005.

Another of Saturn's amazing moons is Enceladus. At 500 km across it is just a tenth the size of Titan and could fit within the island of Great Britain, but it holds many amazing secrets. It is covered in fresh, clean ice making it the most reflective body in the Solar System. Enceladus looks like a giant snowball!

Slicing through this pristine covering are four blue-green bands, 130 km long, known as the 'tiger stripes'. Geysers of liquid water erupt thousands of kilometres into space where they freeze into snowy crystals that then fall on the moon's surface. Some of the icy material drifts off to help form one of Saturn's outer rings. With a liquid water ocean beneath its surface, internal heat and complex hydrocarbons, Enceladus is another favourite place of astronomers looking for signs of extraterrestrial life.

SATURN'S MOONS

Most of Saturn's known sixty-two moons are quite small. Only thirteen have diameters larger than 50 km and the majority are irregular in shape rather than spherical. The most notable exception, Titan, orbits 1.2 million km away from Saturn and is the second largest moon in the Solar System. Already known to be the only moon in the Solar System to have an atmosphere, the Cassini-Huygens space probe also discovered that it is the only known place outside Earth to have stable surface lakes. Ranging in size from 1 km to 100 km wide, they are pools of liquid ethane, methane and dissolved nitrogen.

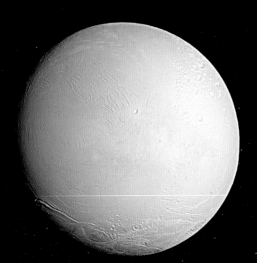

Enceladus appears to have an icy surface with liquid water underneath.

This image shows the giant moon Titan behind Saturn's rings and the tiny moon Epimetheus, the fifth in distance from Saturn

Hyperion is the largest irregularly shaped moon ever observed. Potato-shaped, it has an average diameter of 270 km.

Mimas is believed to be composed of water-ice and has a distinctive 130 km-wide crater almost one-third of its diameter.

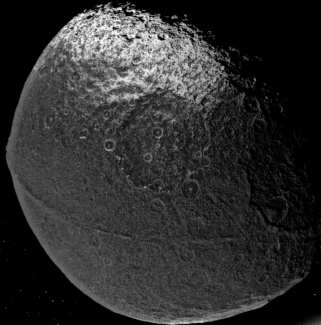

Iapetus has a mountain ridge over 20 km high that stretches around its equator.

Phoebe orbits Saturn in the opposite direction to most other moons.

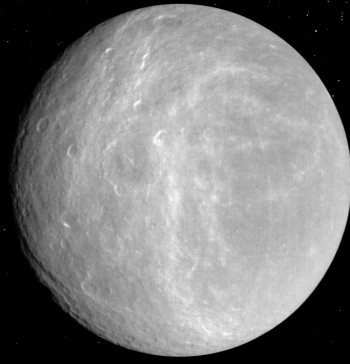

Rhea is Saturn's second-largest moon.

Titan is Saturn's largest moon and the second-largest in the Solar System. A dense orange smog cloaks a surface topography with methane rivers and lakes.

SATURN — FACTS

1427 million km

Distance from Sun

29 Earth years

Length of year

10 hours 39 minutes

Length of day

116 500 km

Diameter

366 000 km

Circumference

-178 °C

Average temperature

62

Number of moons

The bright feature on this image is a storm on Saturn's surface
with an estimated speed of over 1 500 km per hour.

Saturn is more than twice the distance away from Earth as Jupiter

Earth

Jupiter

Saturn

629 million km

1277 million km

116 500 km

270 000 km

Saturn's rings are more than twice the planet's diameter

URANUS

Uranus is the seventh planet from the Sun. It was first identified as a planet in 1781 by the British astronomer William Herschel and is named after the Greek god of the sky. It has the third-largest planetary diameter and fourth-largest planetary mass in the Solar System.

Uranus is different in its composition to the larger gas giants Jupiter and Saturn. Its atmosphere is also mostly made up of hydrogen and helium, but it also contains more 'ices' such as water, ammonia, and methane, along with traces of hydrocarbons. Astronomers sometimes refer to it and Neptune as 'ice giants'.

When Voyager 2 flew past Uranus in 1986 it sent back pictures of what seemed to be a rather featureless blue planet. But Uranus has plenty of surprises in store. With temperatures plummeting to −224 °C, Uranus has the coldest planetary atmosphere in the Solar System. Its cloudy atmosphere is built of swirling layers, with water making up the lowest clouds and methane on top. Methane absorbs red light, making the planet seem pale blue.

You couldn't land on Uranus – not only do wind speeds in the atmosphere reach 900 km/h, but there is no solid surface just poisonous gas that gets progressively denser until it becomes a liquid. The very heart of the planet is mainly built of ices and rock.

One of the most amazing things about Uranus is the way it spins. The planet is tipped right over on its side, with its north and south poles lying where other planets have their equators. So where all the other planets rotate like tilted spinning tops, Uranus rotates more like a tilted rolling ball. It also spins on its axis in the opposite direction to most of the other planets. This is called retrograde. Uranus's unusual spin was probably caused by another planet-sized object smacking into it millions of years ago and knocking it over onto its side.

This oddly tilted axis creates extreme seasons. At Uranian solstices, one pole faces the Sun continuously while the other points away. Each pole bathes in 42 years of continuous sunlight, followed by 42 years of utter darkness. Around its equinoxes, Uranus's equator faces the Sun, resulting in day-night cycles similar to those on other planets.

Thirteen distinct rings encircle Uranus. They are thin and made up of mostly coin-sized particles too faint to see from Earth. The planet also has twenty seven known moons. Titania is the largest at 1,610 km in diameter. Miranda is one of the oddest in the Solar System. Scarred with huge ridges, craters, and canyons twelve times deeper than the Grand Canyon and 80 km wide, it was smashed into pieces by a collision before gradually reassembling. It's the Frankenstein's monster of moons!

URANUS — THE BLUE PLANET

Lying at twice the distance from the Sun as Saturn, the discovery of Uranus in 1781 immediately doubled our perceived size of the Solar System. It was the first planet to be discovered through a telescope although at its brightest it can just be seen as a faint light with the naked eye from Earth. Most of our knowledge of the planet comes from the Voyager 2 flyby in 1986. This has been the only spacecraft to visit Uranus and the images it returned showed the southern pole facing the Sun, illustrating that Uranus is the only one of the Gas Giants tilted so that its equator is nearly at right angles to its orbit.

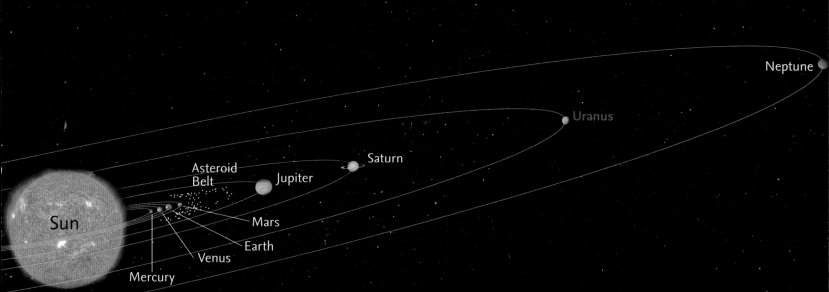

Neptune

Uranus

Saturn

Asteroid
Belt

Jupiter

Sun

Mars

Earth

Venus

Mercury

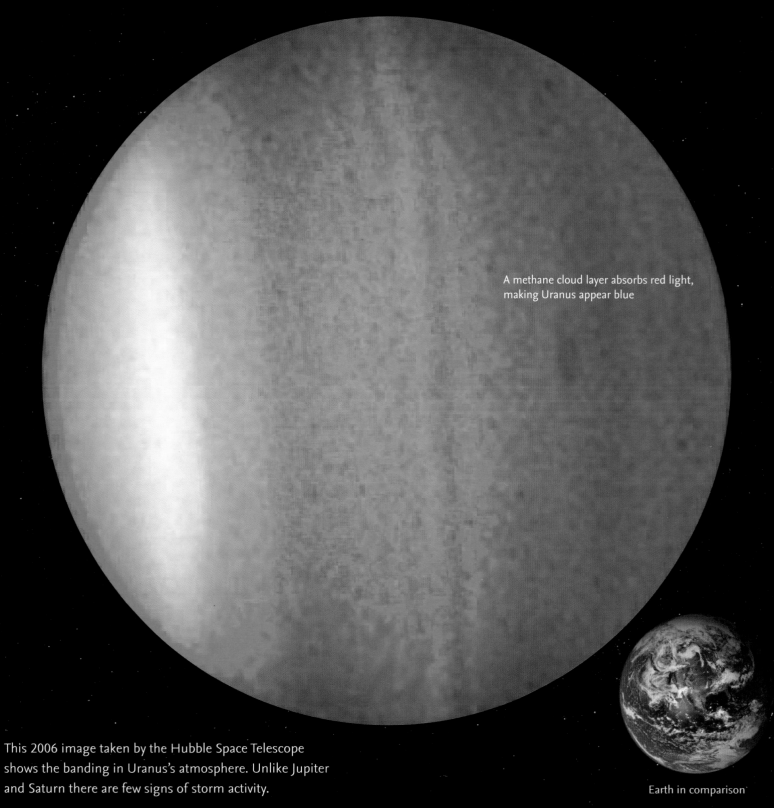

A methane cloud layer absorbs red light, making Uranus appear blue

This 2006 image taken by the Hubble Space Telescope shows the banding in Uranus's atmosphere. Unlike Jupiter and Saturn there are few signs of storm activity.

Earth in comparison

URANUS – FACTS

2871 million km

Distance from Sun

84 Earth years

Length of year

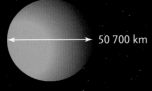

17 hours 14 minutes

Length of day

50 700 km

Diameter

159 000 km

Circumference

-216 °C

Average temperature

27

Number of moons

Miranda
Puck
Ariel
Umbriel
Titania
Oberon

While most planetary satellites in the Solar System take their names from mythology, the six largest moons of Uranus are named after characters from the works of William Shakespeare and Alexander Pope

The extreme seasons on Uranus

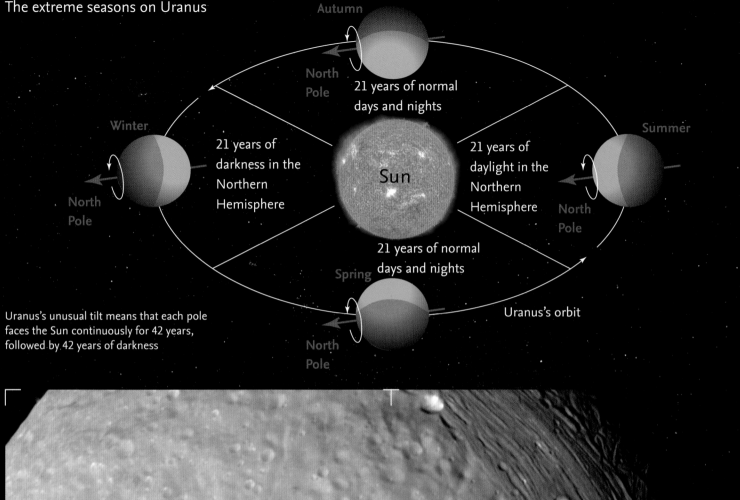

Autumn

North Pole

21 years of normal days and nights

Winter

North Pole

21 years of darkness in the Northern Hemisphere

Sun

21 years of daylight in the Northern Hemisphere

Summer

North Pole

21 years of normal days and nights

Spring

North Pole

Uranus's orbit

Uranus's unusual tilt means that each pole faces the Sun continuously for 42 years, followed by 42 years of darkness

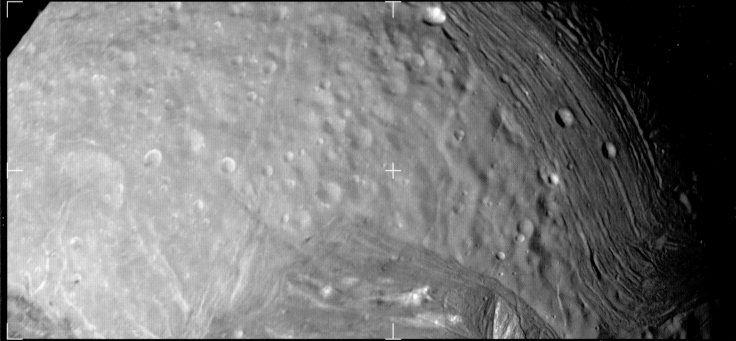

This image of the surface of Miranda, Uranus's fifth largest moon, displays many different geological features with grooves, ridges and markings not seen anywhere else in the Solar System

NEPTUNE

Neptune is the eighth and furthest planet from the Sun. It orbits at thirty times the distance that Earth does and spins on its axis once every sixteen hours. It was named after the Roman god of the seas, probably because of its deep blue colour. Neptune has been visited by one spacecraft, Voyager 2, which flew by in 1989.

The planet was discovered in 1846 by calculation, not observation. Astronomers worked out that a heavy, undiscovered object was tugging at Uranus, messing with its orbit. Neptune was later spotted very close to where mathematical predictions said it would be.

Neptune is more similar in its composition to fellow ice giant Uranus than any other planet. Like Uranus, methane in its atmosphere absorbs red light to make the planet look blue. But Neptune appears a rich azure compared with Uranus's lighter cyan. White wispy clouds of methane ice crystals up to 200 km wide streak across the blue.

Neptune's atmosphere is more feisty and active than its sister planet's. Gales have been clocked at 2,100 km/h, making it the windiest planet in the Solar System. Storms rage through its atmosphere, appearing to us as great swirling spots. One huge tempest measured 13,000 km across, as wide as the Earth. It was known as the Great Dark Spot and was similar to the Great Red Spot on Jupiter. In the 1990s the spot vanished and another large storm appeared in a new location.

Neptune is tilted on its axis in a similar way to Earth and Mars. This means the planet has distinct seasons as it orbits the Sun.

But since it takes 165 Earth years to go once round the Sun, each season on Neptune lasts for 40 of our years!

Orbiting Neptune are at least thirteen moons and four faint rings of icy particles. The largest moon by far is Triton, which accounts for 99.5 per cent of the mass orbiting the planet. The Sun would just look like a bright star from Triton and Neptune would be a huge blue world hanging in the sky. Triton's surface is mostly frozen nitrogen, water ice and carbon dioxide, reflecting much sunlight and making it very cold. It is also one of the few moons that is geologically active, with erupting geysers that spurt out nitrogen.

Triton orbits Neptune in the opposite direction to all the other moons, which suggests that it didn't form with the planet but was later grabbed by its strong gravity and pulled into orbit. In 3.6 billion years, this same gravity will pull Triton too close to Neptune. The moon will be ripped apart and the pieces will scatter round the planet to form new ring systems to rival Saturn.

NEPTUNE — THE FURTHEST PLANET

The outermost planet in the Solar system, Neptune is invisible to the naked eye and is seen by most telescopes as only a small blue disc. Its discovery in 1846 greatly increased the known size of the Solar System, as had the discovery of Uranus sixty-five years previously. Taking 165 Earth years to orbit the Sun, it wasn't until 2011 that Neptune completed its first observed circuit since its discovery. Most of our detailed knowledge of this planet comes from the flyby of Voyager 2 in 1989. Returning spectacular photographs of swirling high-altitude cloud, the Hubble Space Telescope continues to increase our understanding of Neptune. Although appearing a calm blue it is the windiest planet in the Solar System. Several dark spots have been seen that indicate great storms. The largest was about the size of Earth and was known as the Great Dark Spot, similar to the Great Red Spot on Jupiter. In the 1990s it vanished and another appeared in a new location, indicating the dynamic weather that dominates this planet.

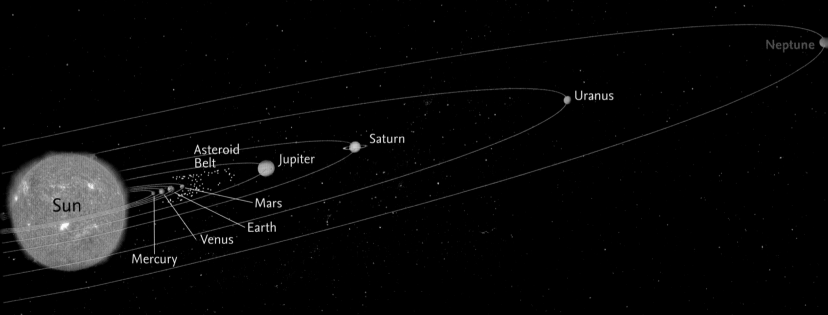

Neptune's deep blue colour is caused by methane in its atmosphere

Voyager 2 observed this triangular-shaped patch of storm clouds called Scooter circling the planet at an even faster speed than the Great Dark Spot

Neptune's most striking feature, the Great Dark Spot was an enormous storm cloud that rotated anti-clockwise and moved westward across the surface of Neptune at a speed of 300 m per second. It was discovered by Voyager 2 but in 1994 was no longer visible to the Hubble Space Telescope

The Small Dark Spot, or Wizard's Eye, was the second fastest storm seen by Voyager 2. Like the Great Dark Spot it too had disappeared by 1994

This image of Neptune was captured by Voyager 2 in 1989. Neptune is the last of the Gas Giants in the Solar System and is more than thirty times as far from the Sun as Earth.

Earth in comparison

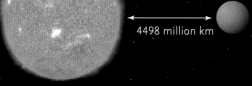

4498 million km

Distance from Sun

165 Earth years

Length of year

16 hours 7 minutes

Length of day

49 200 km

Diameter

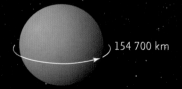

154 700 km

Circumference

-214 °C

Average temperature

13

Number of moons

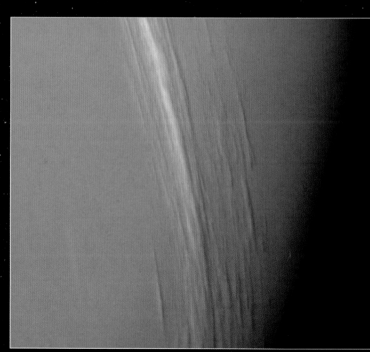

Wispy white clouds on Neptune look like cirrus clouds on Earth but are made of methane crystals. The width of these streaks range from 50 km to 200 km. Neptune is the only planet other than Earth where cloud shadows have been seen

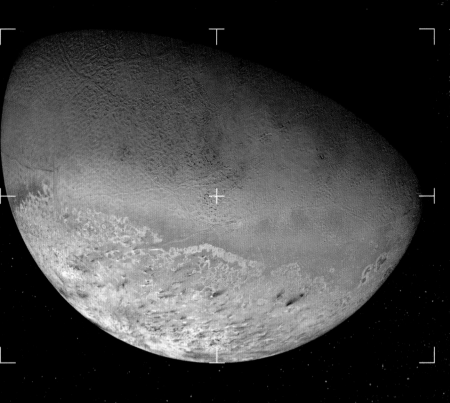

Triton, Neptune's largest moon, is one of the coldest known places in the Solar System, with a surface temperature of about −235 °C. Frozen nitrogen and methane create pink and bluish areas of surface ice and volcanoes of liquid nitrogen erupt, streaking the surface with dust deposits seen as small dark patches in this photomosaic created by Voyager 2

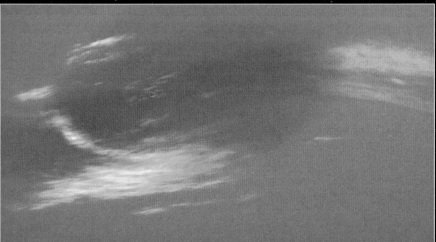

The winds near the Great Dark Spot on Neptune were measured to be about 2,400 km/h and are believed to be the fastest found so far in the Solar System

Neptune is thirty times the distance away from the Sun than Earth

Sun

Earth

Neptune

150 million km

4498 million km

PLUTO

When 24-year-old Clyde Tombaugh discovered Pluto in 1930 it was hailed as the ninth and furthest planet from the Sun. But as more dwarf planets were discovered, astronomers realized that Pluto was only one of several large bodies in the Kuiper belt. Furthermore, Eris, a trans-Neptunian object discovered in 2005, is even larger. Astronomers began to debate what Pluto should be classed as, and in 2006 they decided it should be known as a dwarf planet.

Pluto is a long way away, orbiting forty times further out from the Sun than the Earth. Its year lasts 248 Earth years. It is a small world, with a diameter of 2,322 km it is only two-thirds the diameter of our Moon. It is so far away and so difficult to photograph that studying it from Earth is equivalent to reading the maker's name on a football placed 64 km away.

It is a rocky icy world, freezing in the darkness at a temperature of -233 °C. Pluto's rocky surface is mostly covered in nitrogen ice with some methane and carbon dioxide. It has a thin atmosphere of these same gases, released from the ices on its surface. Pluto was named by Venetia Burney, an 11-year-old girl from Oxford, after the god of the underworld – fitting for such a cold, dark planet!

Five moons orbit Pluto. Charon, the largest, was discovered in 1978 and is more than half the size of Pluto. The two objects spin around each other, taking six days to complete a full orbit. They are tidally locked, so if you were on Pluto, the moon would appear to hang still in the sky. It is such a close relationship that some astronomers consider Pluto and Charon to be a dwarf double planet.

The paths of the other planets around the Sun are mostly circular and lie close to the ecliptic plane. Pluto's orbit is very different. Its path is highly elliptical and sometimes swoops inside Neptune's orbit. It is more tilted too, rising far above the ecliptic. This takes it out of Neptune's path, so the two will never collide. Like Uranus, Pluto's axis is tilted, so it spins almost on its side.

As the only planet never visited by a spacecraft, Pluto is a mysterious world. It will keep many of its secrets until the NASA's New Horizons mission arrives in 2015. New Horizons will photograph the surfaces of Pluto and Charon and investigate their geology. Some of the ashes of Clyde Tombaugh, who died in 1997, are on board the spacecraft. So, in a way he will be the first to visit the dwarf planet he discovered!

PLUTO — THE DWARF PLANET

It is only a small dot of light in even the largest of Earth's telescopes, so Pluto is still very much a mystery. It has been classified as a plutoid, a dwarf planet that orbits the Sun beyond Neptune and is large enough to be rounded in shape by its own gravity. Only four plutoids have been designated, of which Pluto is the second largest after Eris. Although Pluto's largest moon, Charon, was discovered in 1978 its four smaller moons weren't found until 2005 (Nix and Hydra), 2011 (P4) and 2012 (P5). All four were first spotted by the Hubble Space Telescope. Scientists believe that Pluto could have many more moons, and also that there may be many more dwarf planets, all awaiting discovery.

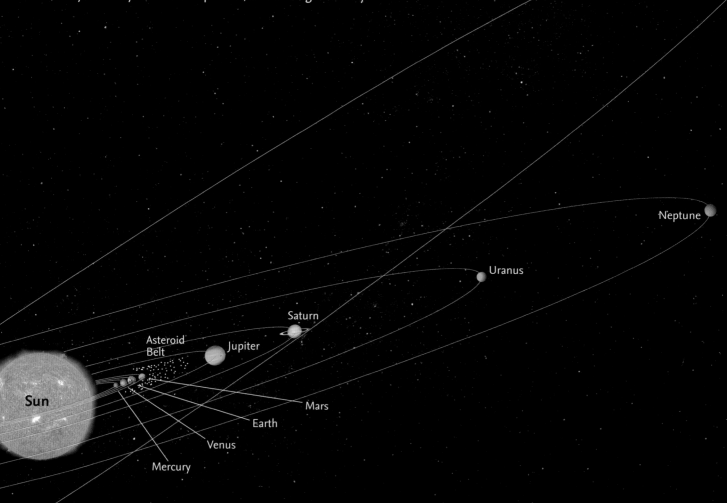

Pluto

Neptune

Uranus

Saturn

Asteroid
Belt

Jupiter

Sun

Mars

Earth

Venus

Mercury

Pluto's has darker and lighter regions across its surface. Until NASA's New Horizons spacecraft arrives in 2015 the reasons for these variations will remain a mystery although the changing distribution of frosts is one likely explanation

Pluto is so small and distant that the task of seeing its surface from Earth is as challenging as trying to see the markings on a football 64 km away. It is known to have surface ice including frozen methane, carbon dioxide and nitrogen.

Pluto and Earth compared

COMETS

Comets can be the most exciting of all space spectacles. Like a continually bursting firework they blaze a trail across the sky, some so brightly that they can be seen during the day. It seems hard to believe that they are really dirty lumps of ice and rock.

The word 'comet' comes from the Greek *kome*, meaning hair: Aristotle thought they looked like stars with long hair. They spend most of their lives uneventfully in the cold dark depths of the outer Solar System. Many have been drifting unchanged in the Oort cloud, 1,000 billion km from the Sun, for 4.6 billion years, making them some of the oldest objects in the Solar System.

But occasionally, one of these dirty snowballs is knocked inwards by the gravitational influence of a planet.

It then sets off on a long, looping orbit towards the Sun. And when it reaches the stage of the inner Solar System, it puts on a wonderful, if brief, show.

The Sun warms the solid part of the comet, the nucleus. Ice is vaporized into powerful jets that shoot from the nucleus, creating a mini-atmosphere around the comet, called a coma. The solar wind then pushes this coma into a long shining tail of gas and dust that can stretch for 100 million km away from the Sun. And so the great firework is born.

When comets visit they leave behind long trails of dust particles. As the Earth passes through these it can create a spectacular meteor shower. Hundreds of bright shooting stars per hour often rain across the night sky.

The Perseids in August and the Geminids in December put on particularly good shows.

Short-period comets take less than 200 years to orbit the Sun and come from the Kuiper belt, beyond the orbit of Neptune. Long-period comets take anything from 200 to thousands or even millions of years to orbit the Sun and come from the Oort cloud, in the outer Solar System.

Comets were once regarded as omens, or signs from God. When a comet appeared in 1066 at the time of the Norman conquest of England, it was featured in the Bayeux Tapestry. In 1705, Edmund Halley showed that this was the same body that had appeared in 1531, 1607, and 1682. He successfully predicted its return in 1759, and it is now known as Halley's Comet.

Comets can meet spectacular ends, often crashing into the Sun or a planet. In 1994 Comet Shoemaker–Levy 9 shattered into pieces and smashed into Jupiter.

COMETS

Comets have oval shaped orbits and only come near to the Sun for a very short time. The solid part, or nucleus, of a comet is surrounded by glowing gases, the coma, which stretches out into a tail. Some comets have more than one tail, a bluish one which trails behind the comet and a yellowish one which follows the path of the comets orbit. Every few years a comet appears that is bright enough to see with the naked eye. Halley's Comet is probably the best known of all. A short-period comet, it last appeared in 1986 when photographs from the space probe Giotto were the first pictures ever taken of a comet's nucleus. Visible about every 75 years, and next appearing in 2061, it is the only comet that might appear twice within a human's lifetime. The European Space Agency (ESA) spacecraft Rosetta is the first mission designed to land on a comet. It is due to enter the orbit of the comet 67P/Churyumov-Gerasimenko in 2014 after a ten-year journey.

Parts of a comet

The tail of a comet consists of two separate tails, one made of charged gases and the other of small dust particles that reflect sunlight. Always pointing away from the Sun, often curved, and often many millions of kilometres long, a comet's tail can make a wonderful sight. As the comet gets closer to the Sun the tail increases in length and as it moves away from the Sun it gets shorter

The coma is a cloud of glowing gases that surrounds the nucleus

The nucleus is the solid part of the comet and is made of ice mixed with rock, dust and grit. Each time a comet passes close to the Sun the ice melts and it loses some of its nucleus

Together the nucleus and coma are called the head

The elliptical orbit of a comet around the Sun

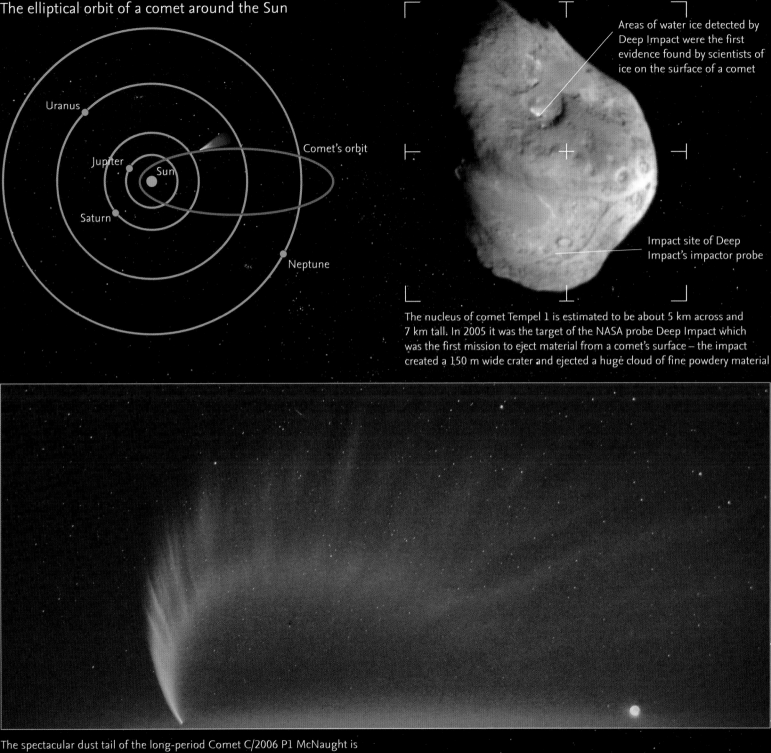

Uranus

Jupiter

Sun

Comet's orbit

Saturn

Neptune

Areas of water ice detected by Deep Impact were the first evidence found by scientists of ice on the surface of a comet

Impact site of Deep Impact's impactor probe

The nucleus of comet Tempel 1 is estimated to be about 5 km across and 7 km tall. In 2005 it was the target of the NASA probe Deep Impact which was the first mission to eject material from a comet's surface – the impact created a 150 m wide crater and ejected a huge cloud of fine powdery material

The spectacular dust tail of the long-period Comet C/2006 P1 McNaught is seen here with the setting sun over the Pacific Ocean on 1 January 2007. At the time it was the brightest comet to be seen from Earth in over 40 years

SOLAR SYSTEM AT A GLANCE

Images not to scale

Sun

Mercury

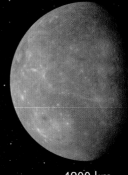

	Sun	Mercury
Diameter	1 391 016 km	4900 km
Circumference	4 370 000 km	15 300 km
Distance from Sun	—	58 million km
Average temperature	5504 °C	-173 °C to 427 °C
Length of year	—	88 Earth days
Length of day	25 Earth days 9 hours	59 Earth days
Number of moons	—	none
Symbol	☉	☿

Jupiter

Saturn

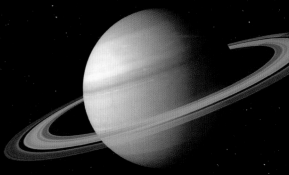

	Jupiter	Saturn
Diameter	140 000 km	116 500 km
Circumference	440 000 km	366 000 km
Distance from Sun	778 million km	1427 million km
Average temperature	-148 °C	-178 °C
Length of year	11 Earth years 314 days	29 Earth years
Length of day	9 hours 55 minutes	10 hours 39 mins
Number of moons	63	62
Symbol	♃	♄

Venus

12 100 km
38 000 km
108 million km
462 °C
224 Earth days 17 hours
243 Earth days
none
♀

Earth

12 700 km
40 000 km
150 million km
15 °C
365 days 6 hours
23 hours 56 mins
1
⊕

Mars

6779 km
21 300 km
228 million km
-63 °C
687 Earth days
24 hours 37 mins
2
♂

Uranus

50 700 km
159 000 km
2871 million km
-216 °C
84 Earth years
17 hours 14 mins
27
⛢

Neptune

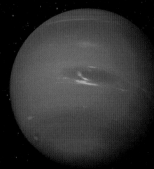

49 200 km
154 700 km
4498 million km
-214 °C
165 Earth years
16 hours 7 mins
13
♆

Pluto

2300 km
7 200 km
4400 to 7300 million km
-233 °C
248 Earth years
6 Earth days 10 hours
4
♇

ASTRONOMY

Ever since humankind first looked up at the night sky and wondered, we have been astronomers. Prehistoric cultures built monuments, such as Stonehenge, to the Sun, stars and planets. Many early civilizations, including the Babylonians, Maya and ancient Greeks, made detailed observations of the stars. This makes astronomy one of the oldest sciences.

But it was with the invention of the telescope in the early seventeenth century that astronomy really came into its own. Galileo built his own telescope and almost immediately discovered Jupiter's moons as well as valleys and mountains on Earth's Moon.

It became clear that the idea that the Earth was the centre of the Universe was wrong. Johannes Kepler was the first to correctly plot the motion of the planets round the Sun. Sir Isaac Newton developed his law of gravitation that finally explained this motion. Newton also developed the reflecting telescope, which proved better than early refracting models.

The fact that Earth's galaxy, the Milky Way, is just one of many such large gatherings of stars was only proved in the twentieth century. Modern astronomy has also discovered many amazing space objects including quasars, pulsars and radio galaxies. These observations have allowed astronomers to develop theories of other exotic objects such as black holes and neutron stars.

Telescopes
Early telescopes were 'refractors' that is, they refracted, or bent the light from stars through lenses. Reflecting telescopes, as perfected by Isaac Newton in 1668, used mirrors to gather the

light. These could be made much larger than refracting telescopes and became the main style of instrument.

In the twentieth century astronomers added more tools to their observing armoury. Radio telescopes can measure 100 m across and be connected together into huge arrays. X-ray and infrared telescopes use wavelengths of electromagnetic radiation outside the visible spectrum. Space telescopes, such as Hubble, avoid the blurring effect caused by Earth's atmosphere, allowing for much clearer observation.

Cosmology

Cosmology focuses on how the Universe began, developed and what will happen to it in the end. Cosmology is studied by physicists and philosophers as well as astronomers.

The most popular cosmological explanation for how the Universe began is the Big Bang theory. This model says that around 13.7 billion years ago all the matter in the Universe erupted from a singularity – a point of incredible density and temperature. This matter, and space itself, rapidly expanded.

The Universe has been expanding ever since and its growth is speeding up. The space between galaxies is growing in all directions. Imagine drawing several dots on a balloon and then blowing more air into the balloon. The dots would all move further apart from each other. Cosmologists believe this is what is happening with the Universe. This expansion may continue forever.

STARS

A star is a shining globe of mostly hydrogen and helium gas that is producing its own heat and light by nuclear fusion. They are constantly being made, with one Sun-like star born in our galaxy at least every year. There are different types of stars, but all are born in a nebula.

Nebulae are vast reservoirs of hydrogen and tiny dust particles. They can be hundreds of light-years across with enough material to make more than a thousand Suns. Look at the middle of Orion's sword with binoculars and you can see a nebula for yourself.

The nebula collapses under gravity, rotating and heating as it does so. This forms a protostar. After 10 million years, if the protostar has at least 8 per cent the mass of our Sun (80 times the mass of the planet Jupiter) it will get hot enough to start nuclear fusion. Hydrogen is converted to helium and energy released. A star is truly born. If a contracting nebula doesn't have this critical mass it shines dimly and is called a brown dwarf or a large planet.

Small stars like our Sun may have enough hydrogen fuel to burn for around 10 billion years. When this is used up, the star balloons into a red giant. It then sheds its outer layers, forming a planetary nebula, before finally shrinking to become a white dwarf.

Large stars are at least twenty times as massive as our Sun. Some are 100 times as massive and shine a million times more brightly. Their fate is very different from a Sun-sized star. They live for only a million years or so, before becoming a red supergiant then a supernova and finally either a neutron star or black hole.

Red giant
With diameters between 10 and 100 times that of the Sun, red giants are very bright. But they have a relatively cool surface, just 2,000–3,000 °C. Red supergiants are 1,000 times larger than our Sun and 1 million times more luminous.

White dwarf

The last stage in the life cycle of a Sun-like star this is very small, hot star. White dwarfs have a similar mass to the Sun, but are just 1 per cent of its diameter, around the diameter of the Earth. This makes them very dense. One spoonful of a white dwarf would have a mass of several tonnes. Their surface temperature is 8,000 °C or more, but their luminosity is just 1 per cent of the Sun's.

Supernova

Big stars go out with a bang. A supernova is the explosive death of a star, which briefly burns with the brightness of 10 billion Suns. There are two main types of supernova:

Type I supernovae happen in binary star systems, where gas from one star falls on to a white dwarf, causing it to explode.

Type II supernovae happen when a star at least ten times more massive than the Sun undergoes runaway internal nuclear reactions at the end of its life. The catastrophic explosion leaves behind a neutron star or black hole. The explosion fuses lighter elements into heavier ones, and supernovae are the main source of elements heavier than hydrogen and helium.

Neutron stars

Neutron stars are very dense, usually having a mass of three times the Sun in a sphere just 20 km across. They are so dense that a spoonful would have 900 times the mass of the Great Pyramid. If it is any more massive, the dying star's gravity will be so intense that it will shrink to become a black hole. Pulsars are rapidly spinning neutron stars.

Black holes

Black holes form when massive stars die. Their gravitational pull is so strong that nothing, not even light, can escape from it. Black holes warp the space around them, and can vacuum up nearby matter, including stars.

STARS LIFE CYCLE

A star's fate depends on its initial mass. If it starts small, like our Sun, it will die slowly. First it will grow to a red giant before shedding its outer layers and shrinking to become a white dwarf. If it begins life as a large star, it will die spectacularly, exploding in a supernova leaving a super-dense neutron star or a mysterious black hole.

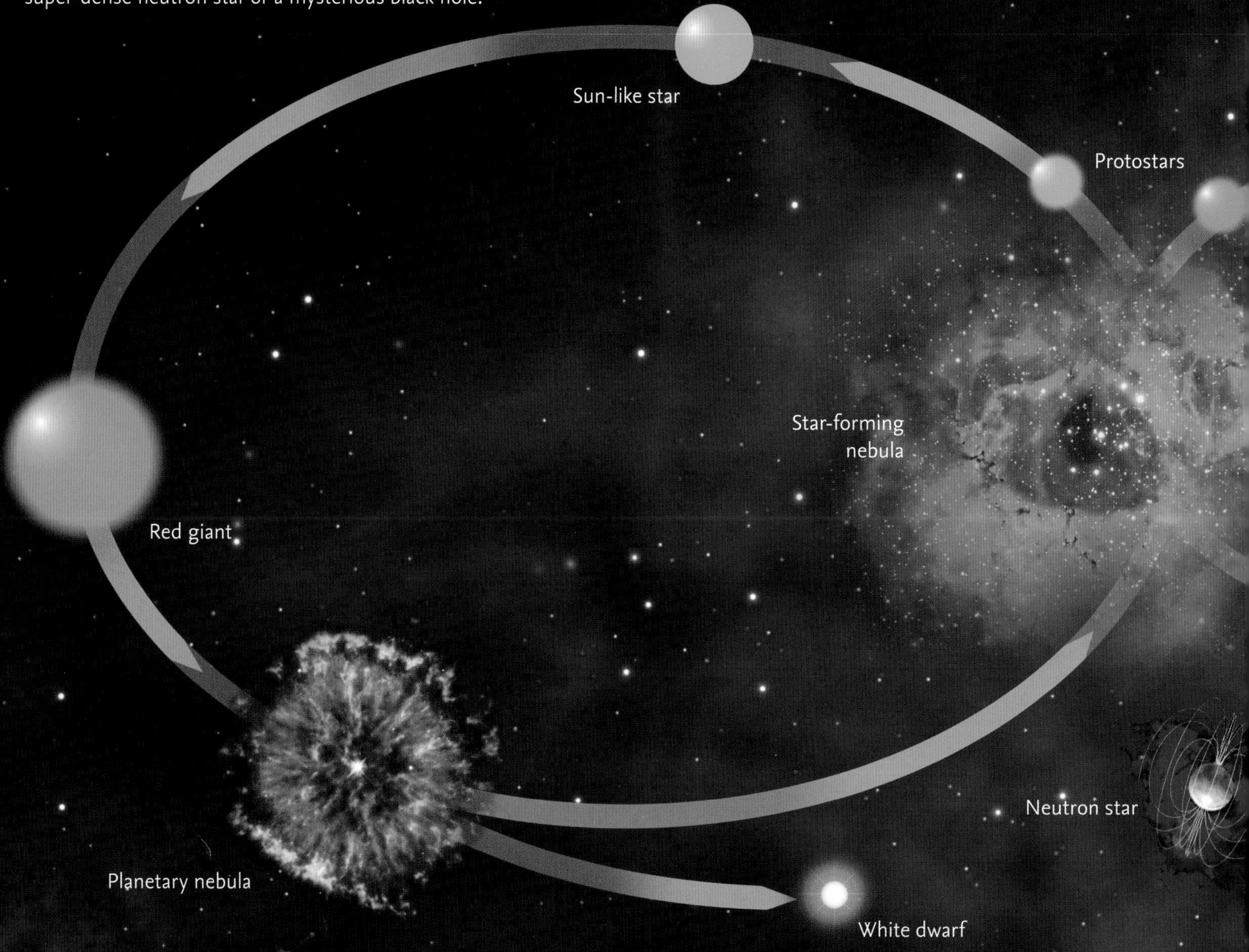

Sun-like star

Protostars

Star-forming nebula

Red giant

Neutron star

Planetary nebula

White dwarf

Massive star

Red supergiant

Supernova

Black hole

This Hubble Space Telescope image shows many elements of the life cycle of a star. The nebula shown is known as NGC 3603. It contains Bok globules – the dark clouds top right, and gaseous pillars at the bottom. A starburst cluster above the centre of the image contains massive stars that are young and hot. At the end of the life cycle is the blue supergiant called Sher 25 with its ring of glowing gas. The blobs to the right and left of it are called bipolar outflows and they show that there is chemically enriched material present.

NEBULAE

Is it a star, a cloud, a planet? A nebula (plural nebulae) can look like any one of these things, but really it is a gigantic cloud of interstellar dust and gases, mostly hydrogen and helium. Nebulae can be light-years across and there are several different types. Several individual nebulae are famous for their beautiful shapes, colours and patterns.

Many nebulae are star nurseries – suns are born and grow in them. This process begins when clumps of denser material form within the gas clouds. The gravity of these clumps can pull in more matter. When the clump gets big and hot enough, it can form a star. If the star has enough energy it can blast the remaining dust and gas in the cloud with ultraviolet radiation. The nebula then appears to be lit up. Matter that doesn't go into the star may form planets, comets and other objects. This is what happened in our Solar System, 4.6 billion years ago. Other nebulae are the remnants of dying stars. These formed when stars exploded in a cataclysmic supernova and threw off a wispy shell of material.

An emission nebula, like the Orion Nebula, glows brightly when gas clouds are energized by the ultraviolet light from stars it has already created. Emission nebulae are usually red or pink and are very, very hot because of the highly energetic new-born stars that bombard their surroundings with radiation.

Ever seen dust particles shining in a sunbeam? A reflection nebula shines in a similar way. Radiation scatters off gas and tiny grains of dust that reflect the energy. The nebula surrounding the Pleiades Cluster is one of this type. Reflection nebulae are usually blue because that's the colour that is most efficiently reflected by the small dust grains.

Dark nebulae form when dense clouds of molecular hydrogen absorb the light from stars behind them. The Horsehead Nebula in Orion is a good example of this, its dark silhouette appears perfectly picked out against the brightness behind.

When a red giant star dies, it jettisons clouds of matter into space, leaving a white dwarf at the centre. This is known as a planetary nebula. But it has nothing at all to do with a planet. The name came from an eighteenth century astronomer who thought that one looked like a planet when viewed through the eyepiece of his small telescope. Our modern images show them to be much more intricate gas clouds.

A supernova remnant is all that remains of the cataclysmic death of a giant or supergiant star. Spectacular clouds remain after the huge explosion.

NEBULAE

The **Cat's Eye Nebula** (NGC 6543) is a planetary nebula in the constellation of Draco, 3,300 light-years away. It was discovered by William Herschel in 1786 and is 1.2 light-years across. Planetary nebulae were given their name because the astronomers who first saw them thought they looked like planets. But really they are complex, beautiful shells of glowing gas cast off by a dying star. This nebula has a very intricate structure built of knots, jets, bubbles and thin arches. You can clearly see at least eleven rings around the central eye. Every 1,500 years the dying star shrugged off a dust shell in a huge pulse of energy. Each of these pulses contained more mass than all of the planets in our Solar System. Now they look like the layers of an onion.

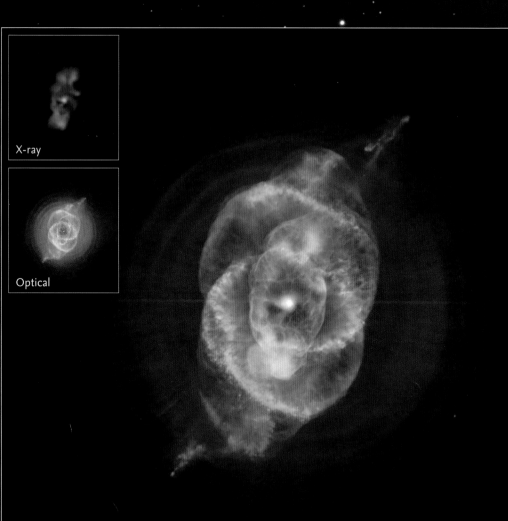

X-ray

Optical

Cat's Eye Nebula, composite

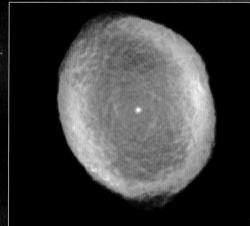

Spirograph Nebula (*Planetary nebula*)
A sculpted jewel with thousands of shining facets, planetary nebula IC 418 lies 2,000 light-years from Earth in the constellation Lepus. Nicknamed the Spirograph after the intricate, looping pictures you can create with the toy, it is 0.2 light-years across. This cosmic gem was a red giant star a few thousand years ago before it ran out of nuclear fuel. Since then its outer envelope has been blown outward, leaving a burning core that will one day become a white dwarf star. Our Sun will die in the same way.

Crab Nebula (*Supernova remnant*)
This supernova remnant forms a truly spectacular nebula in the constellation of Taurus. It was created by a very bright supernova whose explosion was seen by early astronomers in 1054 AD. It is 6,500 light-years from Earth, has a diameter of 11 light-years and is expanding at 1,500 kilometres per second. At its heart is the Crab Pulsar, a neutron star 28–30 km across that rotates 30 times a second.

Cassiopeia A (*Supernova remnant*)
This nebula is a supernova remnant in the constellation Cassiopeia. Although faint visually, it is the brightest radio source beyond our Solar System. The supernova that created 'Cas A' was 11,000 light-years away in our own Milky Way galaxy. The cataclysmic explosion followed the violent collapse of a massive star, probably a red supergiant. At the nebula's core is a black hole.

LBN114.55+00.22 (*Emission nebula*)
This nebula is 6,200 light years from Earth, in the constellation of Cassiopeia. An emission nebula emits light when energy from a nearby star blasts its hydrogen gas, causing it to glow. Dust is also warmed by the light of new stars forming inside the nebula. Since it looks a bit like a chestnut coming out of its shell, perhaps we should call it the Conker Nebula!

Witch Head Nebula (*Reflection nebula*)
You can see why IC 2118 is known as the Witch Head Nebula. Its spooky but elegant shape is an ancient supernova remnant or gas cloud shining with the light of the nearby supergiant star Rigel, in Orion. This makes it a reflection nebula. IC 2118 is in the Eridanus constellation, 900 light-years from Earth. Molecular clouds in the nebula make it a prime spot for new star formation.

Hen 3-1475 (*Protoplanetary nebula*)
Hen 3-1475 is in Sagittarius, 18,000 light-years away. The central star is more than 12,000 times as luminous as our Sun. This hasn't yet blown away its complete shell, so it is a planetary nebula in the making – a protoplanetary nebula. Two S-shaped jets of gas are blasting from this star's poles as it slowly rotates. No wonder astronomers have also nicknamed it the 'Garden-sprinkler Nebula'!

Barnard 68 (*Dark nebula*)
This dark nebula is a huge globule of dust, half a light-year across, that blocks out the light from thousands of stars behind it. The cloud is twice the mass of our Sun and lies within our galaxy, 500 light-years away towards the southern constellation Ophiuchus. Barnard 68 will collapse under gravity within the next 100,000 years or so, beginning its birth as a star.

GALAXIES

Galaxies are huge groups of stars, bound together by gravity. They also contain the clouds of gas where new stars will be born, the remains of dead stars, and a little-understood substance called dark matter. At the heart of every galaxy is probably a supermassive black hole, a star-gobbling monster billions of times as massive as our Sun. Galaxies differ in size. Some dwarf galaxies have as few as 10 million stars while giant galaxies may contain a hundred trillion stars and be a million light years across – ten times the size of our Milky Way.

Galaxies can interact with each other. They can be drawn together into groups of up to fifty galaxies, clusters of several thousand galaxies, and even superclusters where several separate clusters come together. Galaxies can also collide and punch through each other, creating spectacular new hybrids. Elliptical galaxies are often formed from mergers. Our own galaxy, the Milky Way is digesting a couple of smaller galaxies at the moment, and when it collides with the Andromeda Galaxy in 5 billion years, a new elliptical galaxy will be the result.

There are probably around 200 billion galaxies in the observable universe. Astronomers have placed the most distant galaxy ever found at 13.37 billion light-years from Earth. It formed just 380 million years after the Big Bang.

Galaxies also come in different shapes, usually one of three main types: ellipticals, spirals, and barred spirals. Elliptical galaxies have an ellipse-shaped profile that can appear very squashed or nearly circular. Spiral galaxies are disk-shaped with a central bulge of generally older stars. Bright, curving arms sweep outwards from this bulge. Many spiral galaxies have a cigar-shaped band of stars in their core that leads into the spiralling arms. They are known as barred spirals. Our own galaxy, the Milky Way, has this shape. Some galaxies don't easily fit into any of these categories and are called irregular galaxies.

Lenticular galaxies fall halfway between an elliptical and a spiral galaxy, sharing properties of both. Peculiar galaxies have unusual properties caused by other galaxies interacting with them. For example, ring galaxies can form when a smaller galaxy passes through the heart of a spiral galaxy. Starburst galaxies have recently featured a violent event, such as a galactic collision. This has caused a sudden, intense burst of star formation.

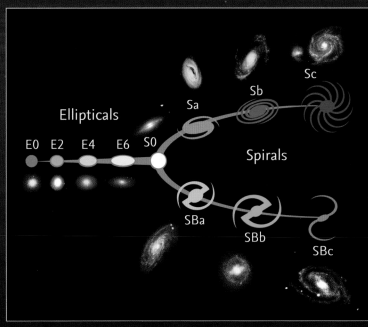

Edwin Hubble studied the galaxies in 1926 and came up with a diagram of their shapes that looked like a musician's tuning fork. It ran from the ellipticals through the hybrid lenticulars and then divided into the two forms of spiral. Irregulars were added at the end of the prongs.

GALAXIES

The Hubble Space Telescope was first launched in 1990. It orbits the Earth once every 97 minutes above the atmosphere to get a better view of the Universe than a ground-based telescope. The telescope's 2.4 m diameter mirror captures light that can then be processed by the various scientific instruments on board. The image below is part of the Hubble eXtreme Deep Field (XDF), which combined ten years of Hubble photographs. It incudes red light captured by the infrared Wide Field Camera 3. You can find out more about Hubble at hubblesite.org

M101 (*Spiral galaxy*)
Also called the Pinwheel Galaxy and NGC 5457, this is a spiral galaxy 21 million light-years away in the constellation Ursa Major. It is 70 per cent larger than the Milky Way, with a diameter of 170,000 light-years. M101 appears lop-sided and this was probably caused by a near collision with another galaxy. This galaxy also has huge and extremely bright star-forming regions. This image of M101 is a composite of pictures from four different NASA space telescopes, using data from the infrared, visible light, ultraviolet and X-ray wavelengths.

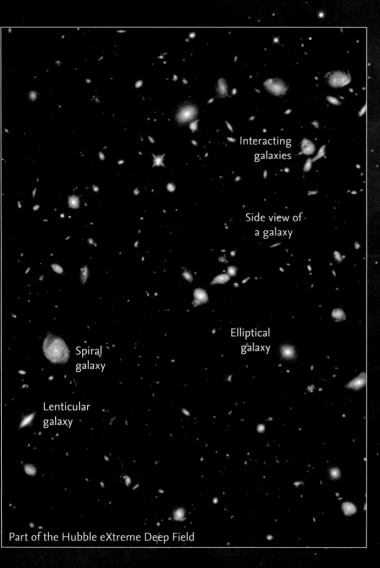

Interacting galaxies

Side view of a galaxy

Elliptical galaxy

Spiral galaxy

Lenticular galaxy

Part of the Hubble eXtreme Deep Field

NGC 1300 (*Barred spiral galaxy*)
NGC 1300 is a barred spiral galaxy about 61 million light-years away in the constellation Eridanus. At 110,000 light-years across this galaxy is slightly larger than our own. It was first spotted in 1835 by William Herschel, the discoverer of Uranus. The galaxy's disk, bulge, arms and core nucleus are clear. You can see blue and red supergiant stars as well as star clusters and star-forming regions in the spiral arms. The nucleus of NGC 1300 has its own distinct structure – a spiral within a spiral. This inner disk is about 3,300 light-years across. Several distant galaxies appear in the background.

NGC 1132 (*Elliptical galaxy*)

An intergalactic pile-up – that's what NGC 1132 is. Galaxy after galaxy slammed into each other, with the resulting collisions creating a giant that outshines a typical galaxy. NGC 1132 still has several dwarf galaxies around it, and together they are called a 'fossil group' because they are all that is left of the ancient galactic pile-up. In visible light NGC 1132 appears as a single large galaxy 240,000 light-years across. But its X-ray glow is ten times that size, the same as an entire group of galaxies.

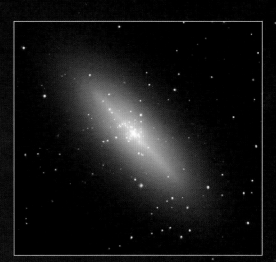

NGC 3115 (*Lenticular galaxy*)

NGC 3115, also called the Spindle Galaxy, is a lenticular galaxy in the constellation Sextans, discovered by William Herschel in 1787. The Spindle is 32 million light-years away from us and many times bigger than our Milky Way. NGC 3115's stars are mostly very old and the galaxy has almost run out of material to make new stars. A supermassive black hole with a mass of 2 billion Suns is believed to be at the centre of this galaxy.

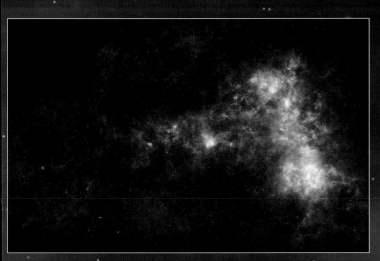

Small Magellanic Cloud (*Irregular dwarf galaxy*)

This dwarf galaxy is one of the Milky Way's nearest neighbours and, at a distance of 200,000 light-years, is just around the cosmic corner. The SMC has a diameter of about 7,000 light-years and contains several hundred million stars. It may have been a barred spiral galaxy that was tugged into an irregular shape by the Milky Way. You can see it with the naked eye in the constellation of Tucana, but only from the Southern Hemisphere. It is linked to the Large Magellanic Cloud.

ESO 593-8 (*Peculiar galaxy*)

Is it a bird, is it a galaxy? This beautiful structure looks like a hummingbird or a cosmic Tinker Bell but is really two massive spiral galaxies and an irregular galaxy crashing together 650 million light-years away. The irregular galaxy is forming stars at a terrific rate.

CONSTELLATIONS

Looking up at the night sky it's easy to imagine that groups of stars form pictures. That's because your human brain is programmed to recognise patterns amongst apparently random dots. Since the earliest times, man has named these groups after animals, figures, gods or other objects. Some likenesses are obvious; others have used quite a bit of artistic licence!

The stars appear to lie on a sphere that surrounds the Earth. This 'celestial sphere' is divided into 88 separate areas of stars, or constellations. The members of each group don't really sit together; the stars themselves may be hundreds of light-years apart.

Many constellations have been called different names by different civilizations. For example, Ursa Major (the Great Bear) was known as Callisto to the ancient Greeks. To Hindus, its seven major stars were seven seers, known as rishis. The Egyptians pictured them as the thigh of a bull, while Europeans saw them as a wagon or a plough.

The constellations appear to move across the sky. This is because the Earth rotates on its axis, and moves around the Sun. Some constellations can only be seen at certain times – Orion in the northern hemisphere winter, and Sagittarius in the northern hemisphere summer for example. Other groups can only be seen from certain parts of the Earth. The Southern Cross will always be invisible to Europeans.

Imagine the Earth inflated to fill the celestial sphere. Just as geographers imagined lines of longitude and latitude on our planet to aid navigation, so astronomers use 'space lines' called right ascension and declination. These are equivalent to longitude and latitude, respectively. Earth's Equator forms an imaginary equator in space. With this grid system astronomers can give any star a set of coordinates, making it easy to find.

Around the celestial equator is a band about 20° wide called the zodiac. Thirteen constellations appear in the zodiac, and astronomers use it as a navigation aid. But this is not the same as the astrological signs of the zodiac. Modern science has found no evidence for the fortune-telling powers of astrology, although in ancient times the movement of the stars and planets through the sky were often believed to be of huge importance to human endeavours. The appearance of a new constellation could signal the change of a season, while a star rising as a child was born could be taken as a powerful omen.

Cataloguing objects

How do astronomers name and locate objects? Space is vast and stars can be very similar. They are listed in several ways. French astronomer Charles Messier loved hunting comets. In 1771 he compiled a list of objects that weren't comets so he didn't have to waste time looking at them! These are known as Messier objects, and there are now 110 on this list.

There are also 7,840 objects with an NGC number. This is from the New General Catalogue of Nebulae and Clusters of Stars compiled by John Louis Emil Dreyer in 1888. His comprehensive list includes all types of deep space objects.

Often space objects are given a home within a particular constellation. This makes it easy to find them quickly.

Northern sky

There are 88 modern constellations. These are all shown on the separate charts for the northern and southern hemispheres on these pages. The key to the chart symbols is in the centre of the spread, top and bottom.

Star Magnitude

Each unit of magnitude indicates a difference in brightness of 2.512 times. The brightest star is Sirus (mag. – 1.43)

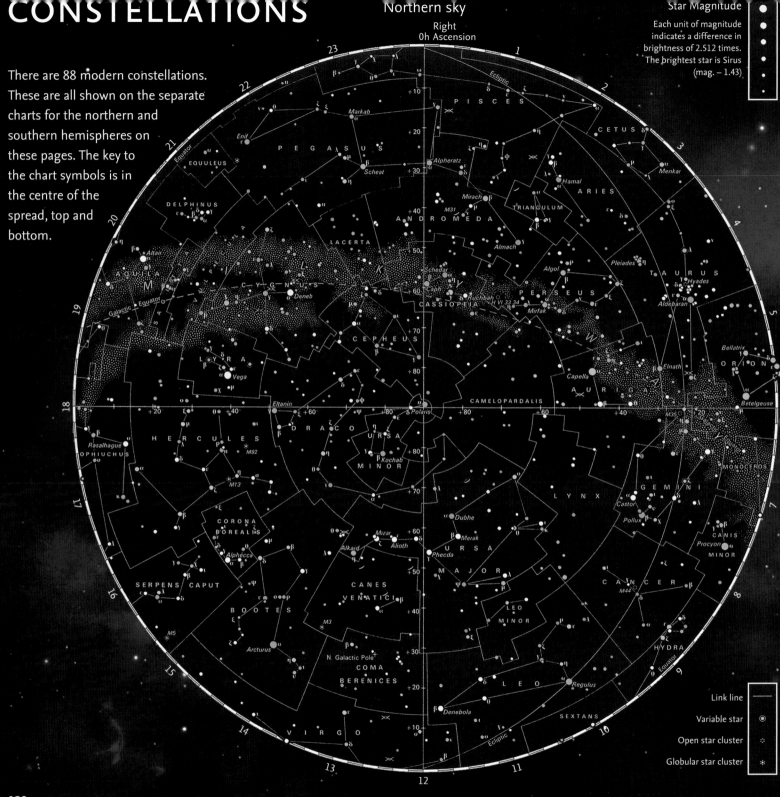

Link line
Variable star
Open star cluster
Globular star cluster

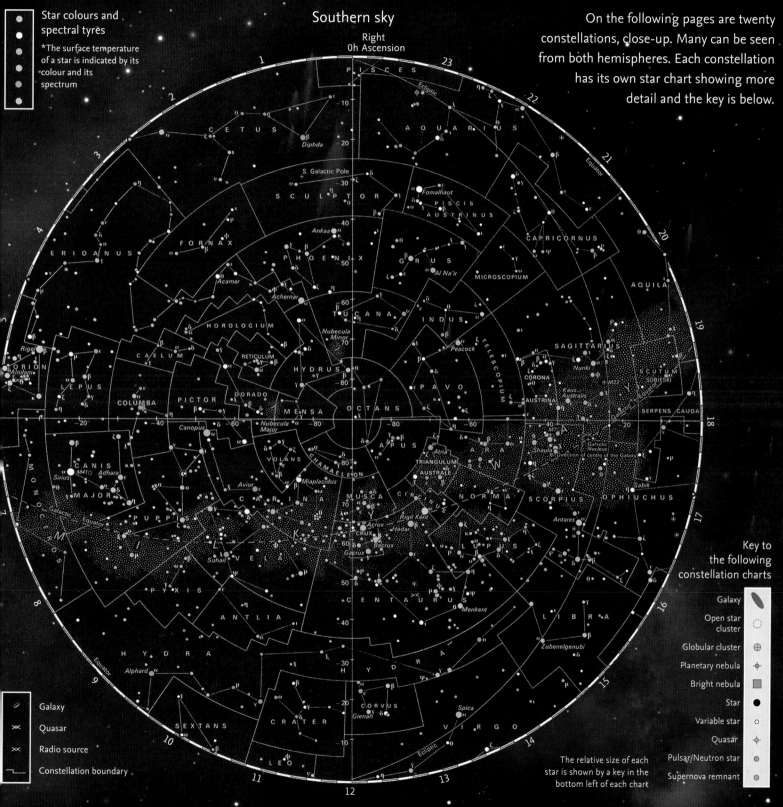

Southern sky

Right
0h Ascension

Star colours and spectral tyres

*The surface temperature of a star is indicated by its colour and its spectrum

On the following pages are twenty constellations, close-up. Many can be seen from both hemispheres. Each constellation has its own star chart showing more detail and the key is below.

Galaxy

Quasar

Radio source

Constellation boundary

Key to the following constellation charts

Galaxy	
Open star cluster	
Globular cluster	
Planetary nebula	
Bright nebula	
Star	
Variable star	
Quasar	
Pulsar/Neutron star	
Supernova remnant	

The relative size of each star is shown by a key in the bottom left of each chart

CONSTELLATIONS CLOSE UP — ANDROMEDA

Location: Northern Hemisphere
Coordinates: Right Ascension 01h, Declination +40°

Andromeda is one of the largest constellations and is named after the daughter of Cassiopeia, who in Greek myth was chained to a rock to be eaten by the sea monster Cetus. Andromeda was saved by the hero Perseus, who used the head of Medusa to turn the monster into stone.

The constellation is in the northern sky, close to several other constellations named after characters in the Perseus myth, and is best spotted during autumn evenings. Its basic shape is an 'A' with a star representing Andromeda's head at the point of the letter. You can also picture her outspread arms chained to the cliff.

Our galactic neighbour

The colossal Andromeda Galaxy is about twice as big as the Milky Way but the two have much in common. They are barred spiral galaxies which have several hundred billion stars. Both are orbited by other smaller galaxies. Indeed, both Andromeda and the Milky Way belong to a cluster of about 50 galaxies known as the Local Group.

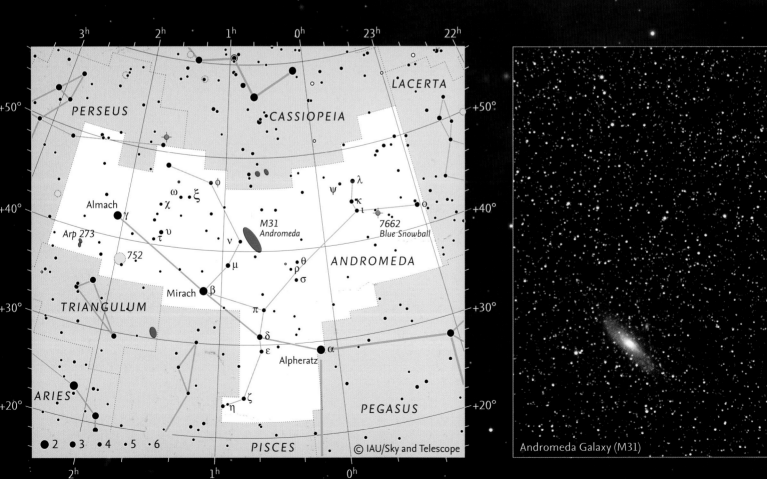

© IAU/Sky and Telescope

Andromeda Galaxy (M31)

Andromeda Galaxy (M31)

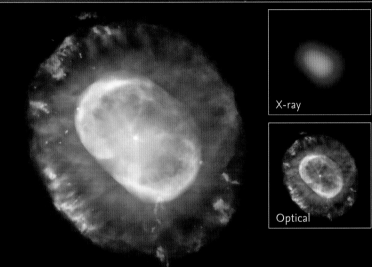

Blue Snowball Nebula (NGC 7662), composite

X-ray

Optical

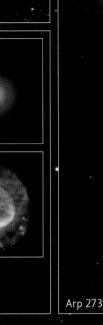

Arp 273

Andromeda and the Milky Way are currently 2.5 million light-years apart. But they are closing on each other at 470,000 km/h. In around 5 billion years they may well collide to form an elliptical galaxy. This may spark a flurry of star-formation, called a starburst. As the closest spiral galaxy to our Milky Way, Andromeda is an ideal target for both amateur and professional astronomers. It is the most distant object you can see with the naked eye. Through binoculars you can clearly see its elongated shape, a result of its tilt relative to our line of sight.

The Blue Snowball Nebula
This planetary nebula (NGC 7662) appears as a beautiful disk with a faint blue central star. The core dwarf star varies in its brightness and is around 5,600 light-years away

Blooming beautiful
Looking like a rose blooming from a twisting stem, or a genie swirling from a lamp, Arp 273 is a group of galaxies 300 million light-years away in Andromeda. The flower is UGC 1810, a large spiral galaxy with a disk that has been twisted by the gravitational tug of the galaxy below it. The magical arc of blue jewels across its top is the light of clusters of super-hot and bright young stars. The smaller galaxy below is known as UGC 1813 and it appears nearly edge-on. New stars are being rapidly made in its nucleus, perhaps triggered by the collision. The smaller galaxy has probably passed right through the larger one. Now the main galaxies are tens of thousands of light-years apart, but are linked by a thin bridge of material.

CONSTELLATIONS CLOSE UP — AQUARIUS

Location: Zodiac constellation in both hemispheres
Coordinates: Right Ascension 23h, Declination -15°

Aquarius is Latin for 'water-carrier' and this constellation has been linked with water for millennia. In many places it rose at the start of the rainy season, making it an important symbol for many cultures.

Aquarius is a large constellation but it doesn't have many particularly bright stars. Excitingly, though, astronomers have recently found that many of its stars have planetary systems.

NGC 7009 Planetary nebula

Does this remind you of a sweet twisted in its wrapper? The

Saturn Nebula (also known as NGC 7009 or Caldwell 55) is a beautiful planetary nebula that shines very brightly. It gets its name because the astronomers who first spotted it thought it looked like the ringed planet Saturn. This nebula is very complex, with a halo, shooting jets, bright gas shells, thin threads and twisted knots. At its centre is a very hot bluish dwarf star that ejected the nebula we see today.

Lyman-alpha blobs

Lyman-alpha blobs (LAB) are among the biggest single things in the Universe. They are vast balloons of hydrogen gas more than 400,000 light-years across. Using the Chandra X-ray Observatory, astronomers have discovered supermassive black holes growing in some blobs. The blobs are lit up by energy thrown out as material

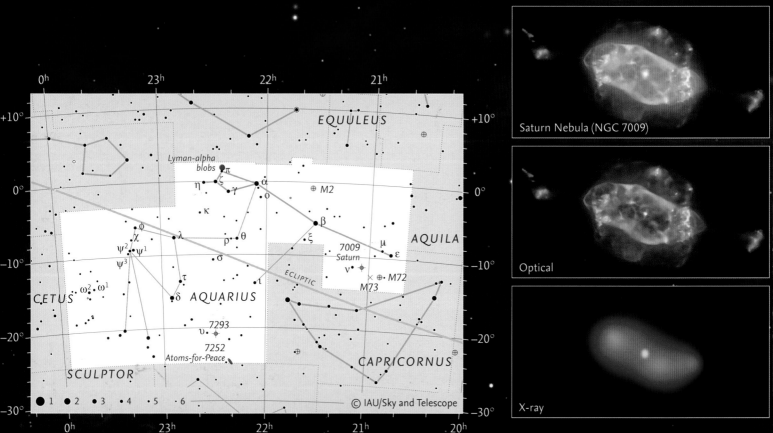

Saturn Nebula (NGC 7009)

Optical

X-ray

© IAU/Sky and Telescope

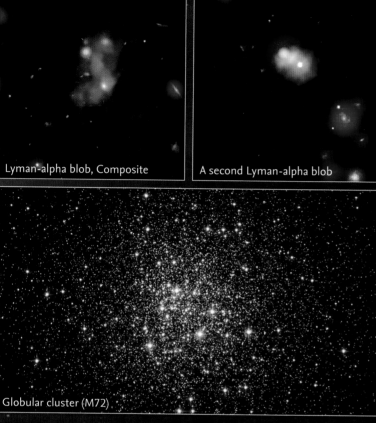

Lyman-alpha blob, Composite

A second Lyman-alpha blob

Globular cluster (M72)

Atoms-for-Peace (NGC 7252)

...alls into the black holes, and by newly formed stars, too. This is a vital time in the birth of a galaxy, and Lyman-alpha blobs may help us understand how galaxies grow.

M72 Globular cluster

Globular clusters were the kings of the young Milky Way. They roamed our early galaxy in their thousands, but now there are less than 200 left. Many were wiped out when they crashed into each other. The ones that are left are the oldest structures in the galaxy. No more will form now as conditions are not right.

This Hubble Space Telescope picture shows around 100,000 of the stars in cluster M72. This grouping spans 50 light years and lies

Atoms-for-Peace

This cosmic car-crash, where two galaxies are smashing into each other, is an excellent opportunity for astronomers to learn about the evolution of the Universe. You can see how the chaotic collision has made huge tails of stars, gas and dust stream out into space. Gas shells torn from the original galaxies are wrapping themselves round their new joint core. The collision has sparked a flurry of star formation with several hundred young star clusters of 50 to 500 million years old.

Astronomers have christened this cosmic pile-up 'Atoms-for-Peace' after a speech about nuclear power made in 1953 by US President Eisenhower. The merging galaxies look like atoms in

Location: Northern Hemisphere

Coordinates: Right Ascension 15h, Declination +30°

Boötes (pronounced 'boo-ooteez') is a constellation in the northern sky that looks a little bit like a kite or ice cream cone. The name comes from the Greek term for herdsman or ploughman. One of the myths about Boötes says that he invented the plough and so was remembered forever in the stars for his ingenuity.

Boötes contains the fourth brightest star in the night sky, Arcturus. This star's name comes from the Greek for 'bear-keeper'.

It is an orange giant 37 light-years from Earth and is twenty-seven times wider than our sun, although its mass is about the same.

One of the most exciting features of Boötes is that many extra-solar planets have been discovered here. In fact, planets have been found around ten of its stars. The star Tau Boötis is orbited by a large planet that was the first extra-solar planet to be observed directly when it was discovered in 1999. Some Boötes stars have multiple planets.

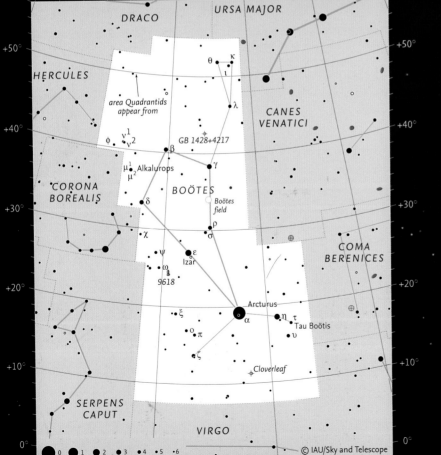

© IAU/Sky and Telescope

Boötes field, X-ray image

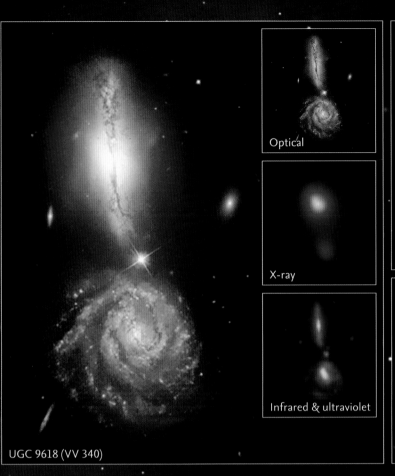

Optical

X-ray

Infrared & ultraviolet

UGC 9618 (VV 340)

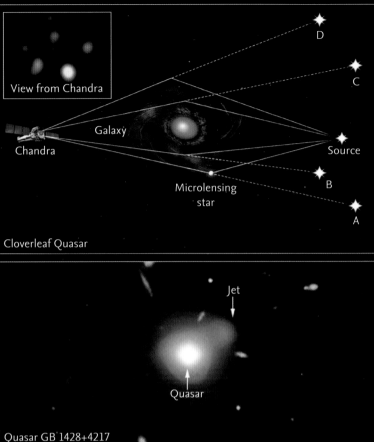

View from Chandra

D

C

Chandra

Galaxy

Source

Microlensing
star

B

A

Cloverleaf Quasar

Jet

Quasar

Quasar GB 1428+4217

Meteor shower

Look up at Boötes on the night of 3 January and you will probably see a fast-moving light-show – the Quadrantid meteor shower. These meteors are faint but there are lots of them – up to 100 zip through Earth's upper atmosphere every hour on this night!

The galactic exclamation mark

UGC 9618, also known as VV 340 or Arp 302, is a stunning pair of spiral galaxies that have recently moved close together. With plenty of gas between them, these galaxies are forming stars at a vigorous rate. In a few million years these spirals will spin into each other and merge. VV 340 throws out vast amounts of infrared light, making it a luminous infrared galaxy (LIRG). LIRGs generate energy hundreds of times more intensely than a typical galaxy. This is probably due to a fast-growing supermassive black hole at the heart of the galaxy.

Cloverleaf Quasar

This is actually one object seen four times thanks to gravitational lensing. This is when the gravitational field of a nearer galaxy bends and magnifies the light from the quasar behind it. We then see several images of the source.

Quasar GB 1428+4217

This huge plume is 230,000 light-years long, or about twice the diameter of the entire Milky Way. It is caused by something amazing – a black hole. As the super-gravity of the black hole sucks material in, energy is released and powerful beams blast away from the black hole at nearly the speed of light. This jet is 12.4 billion light-years away, making it the furthest such quasar ever discovered.

CONSTELLATIONS CLOSE UP — CANCER

Location: Visible in the Northern and Southern Hemispheres
Coordinates: Right Ascension 8.7h, Declination +20.2°

Cancer is a small grouping and is the dimmest of the constellations of the zodiac. Its name comes from the Latin for crab. In Greek mythology the constellation was created when Hercules undertook his famous Twelve Labours, a series of incredibly difficult tasks. In his second adventure he was battling against the Hydra when the goddess Hera sent a crab named Cancer to distract him. The crab seized Hercules's toes with his claws, but Hercules crushed the crab. Hera felt sorry for the small creature that she had sent to its doom, so she gave Cancer the crab a place among the stars.

In ancient times Cancer was particularly important to starwatchers. It was where the Sun reached its most northerly position in the sky – a time known as the summer solstice. Today the solstice occurs in the constellation Taurus, around 21 June.

Stars and clusters
Stargazers often train their telescopes towards Cancer to spot Praesepe (M44 or NGC 2632), an important star cluster known as the 'Beehive'. The Beehive has around fifty stars, making it one of the larger open clusters visible – it is three times the size of the full Moon. If you look at it with the naked eye, it appears as a fuzzy region in the sky. But view it through even a small telescope and its multitude of stars appear like swarm of bees. It buzzes 577 light-years from Earth.

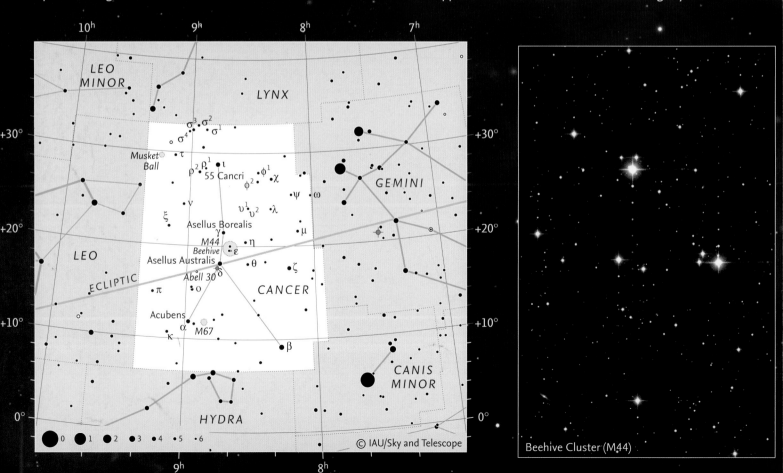

© IAU/Sky and Telescope

Beehive Cluster (M44)

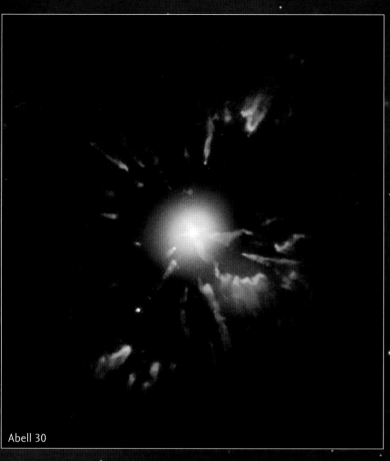

Abell 30

Musket Ball Cluster

The star known as 55 Cancri has a quintuple planet system.
This includes four gas giants and, most excitingly, a 'super-earth'.
This type of planet is more massive than Earth, but not as big
as a gas giant. Super-earths may be good hunting grounds for
extra-terrestrial life.

The born-again nebula

Most stars turn into a nebula at the end of their lives.
Sometimes they do this twice. After burning for several billion
years, Abell 30 used up its fuel and expanded into a red giant,
shrugging off its outer layers. The hot core then blasted the
layers with intense star radiation, creating a planetary nebula – a
colourful shining cloud of gas. The core then repeated its trick,
expanding and turning into a nebula. So a small-scale planetary
nebula was reborn inside the original one – a bit like a gas bubble
inside a previously blown bubble.

Shooting stars

Just as galaxies can crash into each other, so clusters of galaxies
can too. The 'Musket Ball' in Cancer is a system of colliding galaxy
clusters 5.2 billion light years away from Earth. It is similar to the
Bullet Cluster in which astronomers first saw normal matter being
torn apart from the mysterious 'dark matter'. Astronomers gave
the new discovery its nickname because the collision is older and
slower than the Bullet Cluster. Because it is further on in its
evolution, the Musket Ball gives astronomers an excellent insight
into the way clusters change after collisions.

CONSTELLATIONS CLOSE UP – CANIS MAJOR

Location: Northern Hemisphere
Coordinates: Right Ascension 07h, Declination -20°

Canis Major means 'larger dog' in Latin. In Greek and Roman legends this constellation showed one of Orion the hunter's two loyal hounds, the other being Canis Minor (smaller dog). Canis Major contains Sirius, the brightest star in the sky, known as the 'dog star'. Sirius represents the neck of the constellation's dog shape.

Sirius was very important to early civilizations. In ancient Egypt, the dawn rising of the star was a seasonal marker used to predict the annual Nile flood. This signalled the Egyptian new year and the return to the upper world of the Egyptian God of the dead, Osiris. In Roman times, Sirius rose in late July, when the weather became very hot and sultry. This part of the summer was known as the 'dog days' after Sirius's nickname as the Dog Star.

You can easily spot Canis Major because it has many bright stars. Sirius is almost twice as bright as Canopus, the second-brightest star in the sky. It shines so brilliantly because it is very luminous and is just 8.6 light-years away from our solar system. Sirius is also a double star.

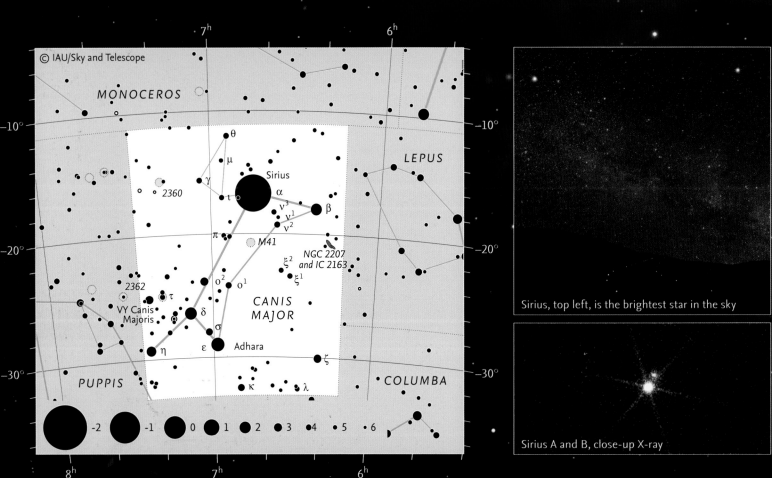

Sirius, top left, is the brightest star in the sky

Sirius A and B, close-up X-ray

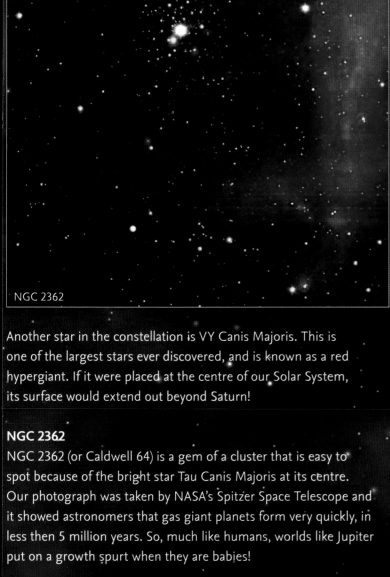

NGC 2362

Hubble image of merging galaxies NGC 2207 and IC 2163

NGC 2207 and IC 2163, Infrared

Another star in the constellation is VY Canis Majoris. This is one of the largest stars ever discovered, and is known as a red hypergiant. If it were placed at the centre of our Solar System, its surface would extend out beyond Saturn!

NGC 2362

NGC 2362 (or Caldwell 64) is a gem of a cluster that is easy to spot because of the bright star Tau Canis Majoris at its centre. Our photograph was taken by NASA's Spitzer Space Telescope and it showed astronomers that gas giant planets form very quickly, in less then 5 million years. So, much like humans, worlds like Jupiter put on a growth spurt when they are babies!

Eyes in the sky

Are you looking at me? This isn't a pair of demonic eyes in a Halloween mask but actually a picture of two merging galaxies, called NGC 2207 and IC 2163, located 140 million light-years away in Canis Major. About 40 million years ago these galaxies drifted together and began dancing a gravitational tango. Swirling in and around each other, their powerful interaction is encouraging new stars to form.

The blue-green eyes in the infrared image are the galactic cores, and the outer mask is made from their intertwined spiral arms that are holding each other as the galaxies whirl across the cosmic dancefloor. The party will eventually end with the pair melding

CONSTELLATIONS CLOSE UP — CASSIOPEIA

Location: Northern Hemisphere
Coordinates: Right Ascension 01h, Declination +60°

Cassiopeia is named after the vain queen in Greek mythology, who boasted that she and her daughter Andromeda were more beautiful than the Nereids, the sea nymphs of Poseidon, the sea god. Poseidon was furious and sent a sea monster to destroy the kingdom. Cassiopeia's husband, King Cepheus, tried to appease Poseidon by sacrificing Andromeda.

The hero Perseus then rescued Andromeda in return for a promise of her hand in marriage. But Cepheus and Cassiopeia went back on this deal once the monster was killed. So Poseidon set images of Cepheus and Cassiopeia in the sky. As a punishment for her treachery, her constellation represents Cassiopeia chained to her throne.

You can easily recognize its distinctive 'W' shape, formed by five bright stars, which is clearest in early November.

Supernova remnants

There are two supernova remnants in Cassiopeia. One is the exploded remains of the supernova called Tycho's Star, which was observed in 1572 by Tycho Brahe.

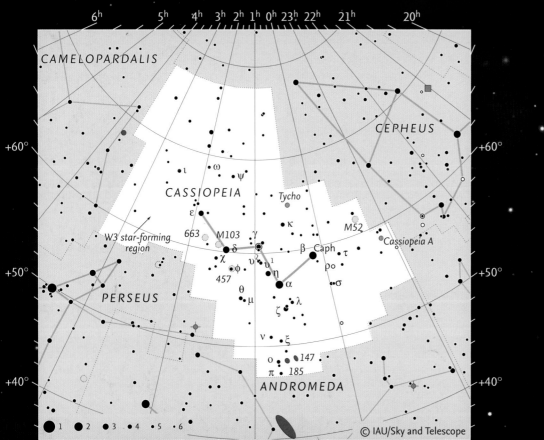

© IAU/Sky and Telescope

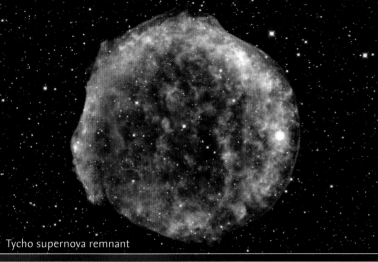
Tycho supernova remnant

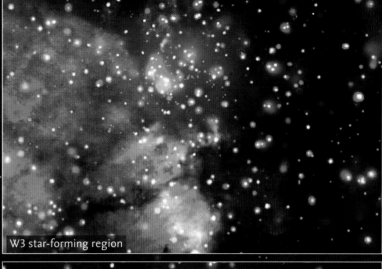
W3 star-forming region

M103 open cluster

X-ray image of W3

The second is the remnant of a supernova that was visible on Earth 300 years ago. The actual explosion happened 10,000 years ago, as the star is light-years away. Known as Cassiopeia A (Cas A), it is the strongest radio source from outside our Solar System. The shell of matter thrown out from the exploded star is moving through space at 4,000 kilometres per second.

M103

M103 (or NGC 581) is an open cluster of about 40 major stars around 8,200 light-years from Earth. Its brightest member is a red giant that is actually a double star. Although it is one of the more remote open clusters, M103 is easy to spot with binoculars.

W3 Star-forming region

W3 is a massive star-nursery where huge new suns are born in clusters. W3 is part of a vast molecular cloud complex. The extraordinary amount of star formation in W3 may have been triggered by a nearby bubble of expanding gas 100 light-years across, called W4. Our picture shows one of the many star-forming complexes of W3. Each bright point is a clutch of several hundred young stars.

CONSTELLATIONS CLOSE UP — CENTAURUS

Location: Southern Hemisphere
Coordinates: Right Ascension 13h, Declination -50°

Centaurus is one of the largest constellations. In Greek mythology, a centaur is a creature that is half man, half horse. Centaurus was the first centaur and the constellation is one of two featuring his kind, the other being Sagittarius.

The constellation is linked with Chiron, the wise, immortal King of the Centaurs. A prophet and skilled healer, Chiron was said to be the first to identify the constellations and teach them to humans. He put a picture of himself in the night sky to guide Jason on his quest for the Golden Fleece.

The grouping forms a large, four-sided shape representing the Centaur's human head and torso, attached to two legs. Rigel Kentaurus is a triple star system that forms one of the Centaur's feet. This system features Alpha Centauri, the brightest star in Centaurus, which is actually a binary star system – two stars revolving round each other. The third, fainter, star is Proxima Centauri. At just 4.21 light-years away, Proxima is the closest star to the Sun, even though it is not visible to the naked eye.

There is an Earth-sized planet in the system, the nearest planet to Earth outside our Solar System. Unfortunately, it orbits its stars ten times closer in than Mercury orbits the Sun, making its surface temperature around 1200 °C – far too hot to be habitable!

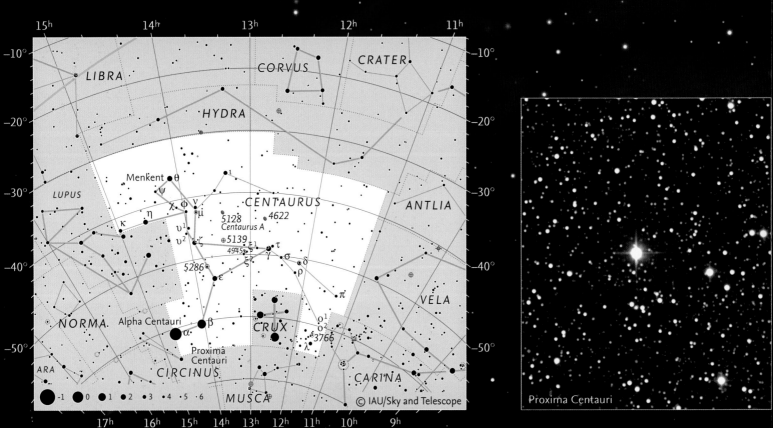

Proxima Centauri

© IAU/Sky and Telescope

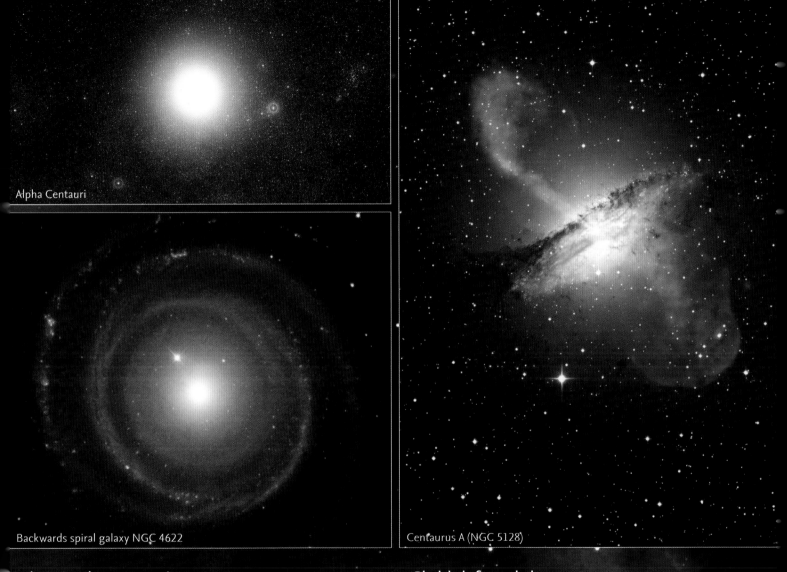

Alpha Centauri

Backwards spiral galaxy NGC 4622

Centaurus A (NGC 5128)

You're going the wrong way!

In most spiral galaxies the swirling arms follow the inner spinning disk, trailing behind its rotation. But spiral galaxy NGC 4622, which lies 111 million light-years away in Centaurus, likes to do things differently. Its spiral arms are leading, with their tips pointing towards the direction of disk rotation. This discovery

Black hole firework show

You can't see a black hole, but you can see spectacular jets and balloons of material powered outwards by its incredible energy. That's what is happening in this beautiful picture of galaxy Centaurus A (NGC 5128). At the heart of this galaxy is a supermassive black hole that is putting on

Location: Southern Hemisphere
Coordinates: Right Ascension 12.5h, Declination -60°

Crux is the smallest of the 88 modern constellations, but if you're in the Southern Hemisphere it is very easy to spot. Its stars are some of the brightest stars of the southern sky and the constellation is a distinct kite shape. The formation is also known as the Southern Cross and is so famous that it even appears on the flags of Australia, New Zealand and Brazil.

The ancient Greeks could see Crux and it was visible as far north as Britain around 6,000 years ago. But the precession of the equinoxes meant that its stars slowly dropped below the European horizon and the people of the north gradually forgot about them.

Dark and bright objects

Crux is home to the Coalsack Nebula, the most easily observable dark nebula in the skies. You can see this as a large dark patch in the southern Milky Way. It is 600 light-years from Earth.

It also holds the open cluster NGC 4755, known as the Jewel Box Cluster. This is about 7,600 light-years from Earth and to the naked eye it appears as a fuzzy star. Through a telescope it appears as a sparklingly beautiful collection of stars, with the

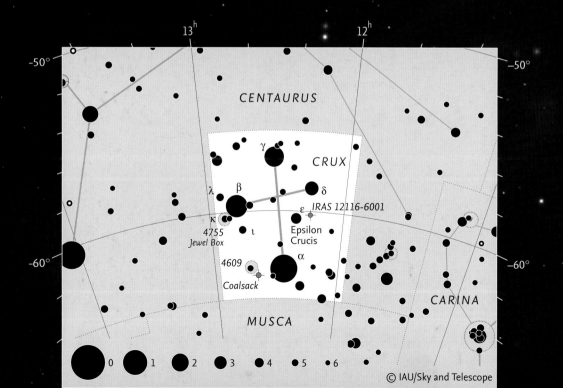

© IAU/Sky and Telescope

Crux (Southern Cross) on the right

Jewel Box Cluster (NGC 4755)

Reflection nebula IRAS 12116-6001

brightest forming the shape of a letter 'A'. The Jewel Box Cluster has more than 100 stars. The brightest are mostly blue supergiants, with a few bright red supergiants.

Seeing in the dark
This stunning cloud of interstellar dust can't be seen directly in visible light, but NASA's Wide-field Infrared Survey Explorer (nicknamed WISE) used infrared to capture this picture of the nebula, called IRAS 12116-6001. What it's actually seeing is clouds of dust heated a little by nearby stars so that they glow with infrared light. Astronomers love reflection nebulae like this because they are prime spots for new star formation.

Stars are sometimes named according to how bright they are within a constellation. The Greek alphabet is used for this, so 'alpha' is the brightest star in the constellation, 'beta' the second brightest, and so on. 'Epsilon' is the fifth letter of the Greek alphabet, so Epsilon Crucis is the fifth brightest star in the constellation Crux.

The different colours in this infrared image represent specific wavelengths of infrared light. Epsilon Crucis appears blue, while the green star nearby is IRAS 12194-6007, a carbon star that is near the end of its life. Because this star is cooler than Epsilon Crucis, the infrared wavelengths it emits are longer.

CONSTELLATIONS CLOSE UP — GEMINI

Location: Zodiac constellation, visible both Northern and
Southern Hemispheres
Coordinates: Right Ascension 07h, Declination +20°

Gemini is linked with the legend of Castor and Pollux from Greek
mythology. Castor and Pollux were identical twins born to Leda,
Queen of Sparta, by two different fathers. Castor was the son of
Leda's husband, King Tyndareus and so was mortal. Pollux's father
was Zeus – the god seduced Leda while disguised as a swan.
Tyndareus adopted Pollux and the twins became inseparable.
When Castor died, Pollux begged his father Zeus to give Castor

immortality, and he did, by uniting them together in the heavens.
A good time to observe Gemini is during January and February.
Look for its two brightest stars Castor and Pollux eastward from
the 'V' of Taurus and the three stars of Orion's belt.

Open and compact clusters

M35 is a relatively scattered, or open, cluster of stars 2,800 light-
years away in Gemini. The cluster's 2,500 stars are mostly young,
just 150 million years old, and are spread over a distance of
30 light-years.

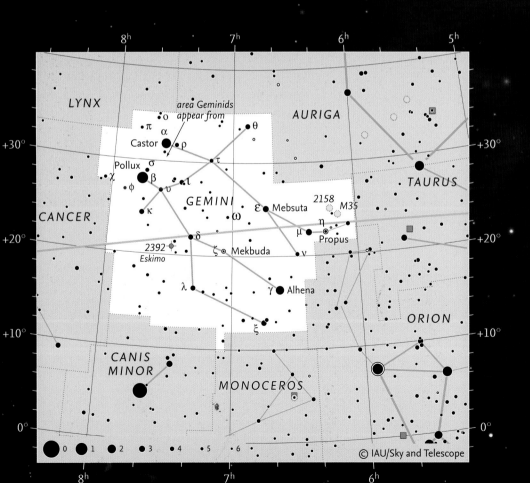

© IAU/Sky and Telescope

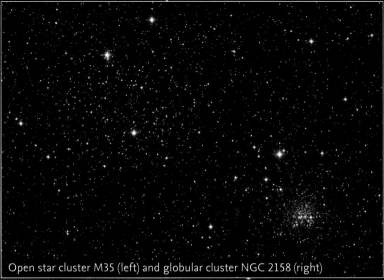
Open star cluster M35 (left) and globular cluster NGC 2158 (right)

The Geminid meteor shower

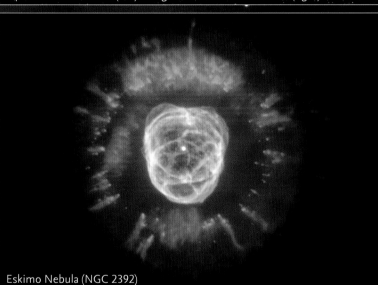
Eskimo Nebula (NGC 2392)

Castor and Pollux, left of centre

The cluster NGC 2158 is ten times older than its neighbour M35 and much more compact, containing a higher number of stars in the same volume of space.

Eskimo Nebula NGC 2392

Looking a little like a pale face in a parka hood, the Eskimo Nebula (NGC 2392) is a planetary nebula with a double shell of ejected gas. It lies more than 2,870 light-years away and was discovered by astronomer William Herschel in 1787.

The nebula's surrounding gas once formed the outer layers of a Sun-like star. The filaments you can see in this outer disk are a light-year long. The inner filaments are blasted out by a strong wind of particles from the central star.

Meteor showers

On 13–14 December the constellation of Gemini puts on a lightshow – the Geminids. This bright meteor shower appears to come from the area of Castor and Pollux. It can top 100 meteors an hour, making it one of the best regular showers of the year.

CONSTELLATIONS CLOSE UP — HYDRA

Location: Northern Hemisphere & Southern Hemisphere
Coordinates: Right Ascension 10h, Declination -20°

It is fitting that the largest constellation should be named after a monstrous serpent. In myth, the Hydra had a dog-like body and many snake heads. One of the heads was immortal. Hydra's breath could destroy life, and her blood was toxic. Fighting the Hydra was the second of Hercules' twelve labours.

He forced the Hydra out of her den with burning arrows and then held his breath while wrestling with her. Every time he cut off a head a new one grew in its place. Eventually he burnt the severed stumps to stop new heads from growing back. He then severed the immortal head and buried it. He dipped his arrows in the Hydra's blood making any wound they caused fatal.

Southern Pinwheel Galaxy (M83, or NGC 5236)

Just 15 million light-years away, M83 is one of the closest and brightest barred spiral galaxies in the sky. It is a star factory, with new suns forming rapidly on a kind of production line. The dark curving 'spines' of the galaxy's arms are dust lanes. New generations of stars grow in clusters on the edges of these lanes.

As these baby stars burst out of their dusty cocoons they produce red bubbles of glowing hydrogen gas.

Eventually, violent winds of charged particles blast the gas away to reveal bright blue star clusters. These stars are up to 10 million years old. The older stars in the galaxy are not as blue. M83 is also Supernova Central, with astronomers spotting the blasted remains of at least 60 exploded stars in this image.

Hydra A galaxy cluster

Galaxy clusters are the largest gravitationally bound objects in the Universe. By observing them, astronomers get vital clues to help them understand the origin and fate of the Universe.

Our image shows two different types of energy coming from a cluster called Hydra A, which lies 840 million light-years from Earth. The pink and blue background represents gas at up to 40 million °C that is emitting X-rays. The green and maroon 'bubbles' are magnetized clouds of radio-emitting particles that have bloomed out from the centre of the cluster, pushing through the X-ray emitting gas.

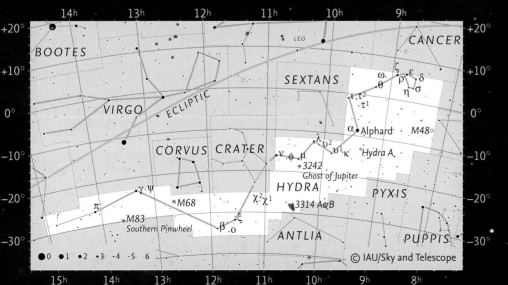

© IAU/Sky and Telescope

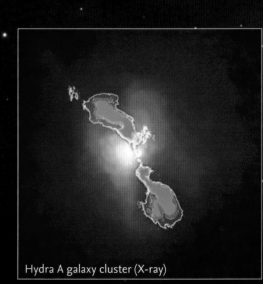

Hydra A galaxy cluster (X-ray)

Southern Pinwheel Galaxy (M83)

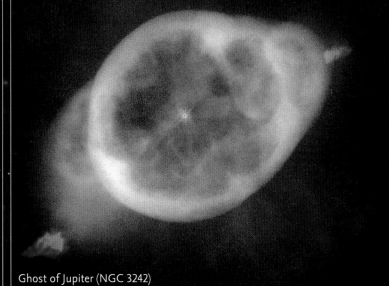

Ghost of Jupiter (NGC 3242)

Close up of Southern Pinwheel Galaxy (M83)

Overlapping galaxies NGC 3314

Ghost of Jupiter

NGC 3242 is a planetary nebula 1,400 light-years from Earth. It is nicknamed 'Ghost of Jupiter' because it looks a little like the giant planet. You can easily see this bluish-green nebula with its white dwarf core through an amateur telescope.

A pair far apart

NGC 3314 is a pair of galaxies that look like they are right on top of each other, but are actually not related or interacting in any way. In the foreground is a spiral galaxy lying face on to us 140 million light-years away. The background galaxy is also a spiral but it is more tilted.

Because they are so neatly lined up, we can see dark material in the front galaxy that would normally be hidden outlined against the glow of the galaxy behind it.

CONSTELLATIONS CLOSE UP — ORION

Location: Celestial Equator (visible in both Northern and Southern Hemispheres)

Coordinates: Right Ascension 05h, Declination +5°

A hunter with broad shoulders, narrow belt and a weapon – so this constellation has appeared to people from the earliest times onwards. The figure is visible everywhere on Earth because it lies on the Celestial Equator. We know it as Orion, a name that comes from Greek stories about a legendary hunter. One myth says that Orion was banished to the sky for boasting about how many

animals he would kill (to impress Eos). He and his hunting dogs, Canis Major and Minor, chase the constellations representing animals, but can never catch them.

Orion is one of the most distinctive constellations. Its seven brightest stars make an hourglass pattern. Four stars mark his shoulders and feet – Betelgeuse, Bellatrix, Rigel, and Saiph. His narrow belt is formed by three bright blue stars – Alnitak, Alnilam and Mintaka. Dangling from this is his sword, a line of two stars and a nebula.

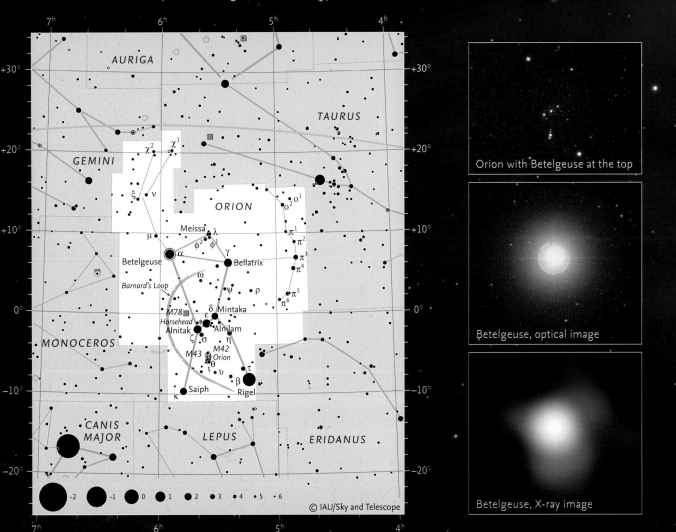

Orion with Betelgeuse at the top

Betelgeuse, optical image

Betelgeuse, X-ray image

© IAU/Sky and Telescope

The stars of Orion's Belt run diagonally across this image, with the Horsehead Nebula and Flame Nebula bottom left

Orion showing Barnard's Loop

Betelgeuse

Betelgeuse is the star at Orion's right shoulder, if he is facing us. A red supergiant near the end of its life, it is one of the largest and most luminous stars in the sky. If Betelgeuse were our Sun, its surface would extend to the orbit of Jupiter, swallowing up Mercury, Venus, Earth and Mars. When it finally explodes within the next million years, the supernova will be so bright it will be visible during the day.

Orion's Belt and sword

Alnitak, Alnilam and Mintaka are blue supergiants, much hotter and more massive than our Sun. They were born in Orion's interstellar clouds and are about 1,500 light-years away.

Orion's sword is made of two stars and the Orion Nebula (M42). Even with the naked eye you can see that this spectacular object is more than a star. With binoculars, you'll glimpse clouds of young stars, shining gas and dust. The Trapezium is a cluster within the

The Horsehead Nebula (or Barnard 33)

You can see why this striking nebula earned its name: it cuts the dark profile of a horse from the swirling pink clouds of ionized hydrogen beyond. The Horsehead Nebula is dark because it is mostly made of thick dust. There are some bright spots in the nebula's base, and these are new-born stars.

Barnard's Loop

The giant Barnard's Loop is an emission nebula 300 light-years across. It is part of the even larger Orion Molecular Cloud Complex, which includes the Horsehead and Orion nebulae. This Complex is one of the most productive stellar nurseries in our galaxy.

The Loop swings in an arc around the Orion Nebula, stretching from the hunter's chest to his toes. It is all that is left of a star that exploded in a supernova 2 million years ago.

tion: Visible in both Hemispheres
dinates: Right Ascension 22h, Declination +20°

sus was the famous winged horse of Greek mythology. He
on of the god Poseidon and the Gorgon Medusa. Pegasus
g from Medusa's body after she was killed by Perseus.

n this magical steed struck the earth with his hoof a pure
g would burst from that spot. On Mount Olympus, Pegasus
iven the task of carrying Zeus' thunderbolts, Zeus thanked
or his faithful service by transforming him into a
ellation.

e are many interesting stars in the Pegasus constellation.
astronomers analysed planet HD 209458b they found the

first-ever evidence for atmospheric water vapour beyond our
Solar System. The planets orbiting star HR 8799 became the first
extra-solar planets to be imaged and the star IK Pegasi is the
nearest likely contender for becoming a supernova.

Stephan's Quintet

This beautiful group of five galaxies, discovered by Édouard
Stephan in 1877, fascinates astronomers because of its violent
collisions. Four of the five galaxies are locked in a gravitational
embrace that will probably end with them merging completely.
One galaxy (NGC 7318B) is falling into the centre of the group at
several millions of miles per hour. As it ploughs into clouds of gas
within the group, a shockwave bigger than the Milky Way spreads
throughout the medium between the galaxies, heating the gas to
millions of degrees.

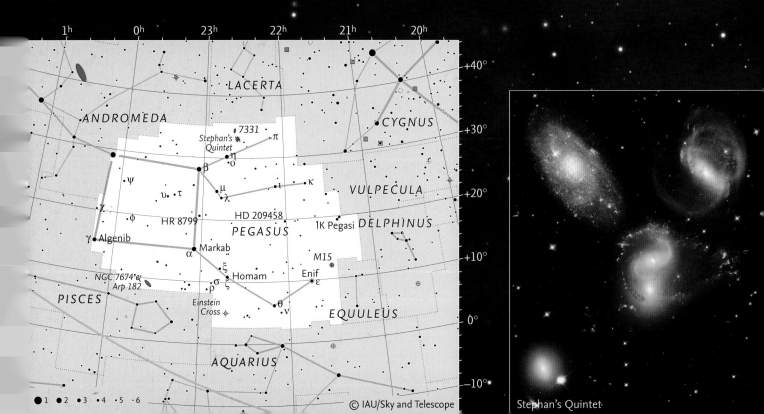

Stephan's Quintet

© IAU/Sky and Telescope

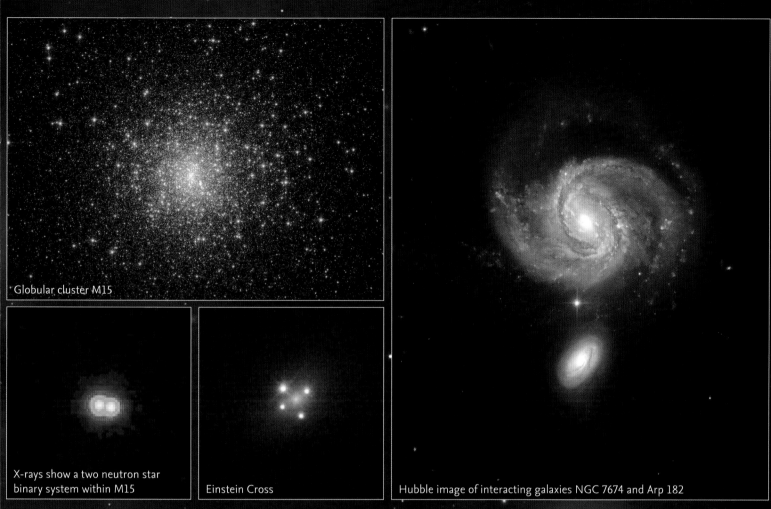

Globular cluster M15

X-rays show a two neutron star binary system within M15

Einstein Cross

Hubble image of interacting galaxies NGC 7674 and Arp 182

M15 Globular cluster

Over 100,000 stars swarm like bees around a massive black hole at the centre of globular cluster M15. Easily visible with binoculars, this ancient relic from the early years of our galaxy is 12 billion years old. M15 is about 33,600 light-years from Earth and 175 light-years in diameter. At its centre is one of the densest concentrations of stars known, with large numbers of variable stars and pulsars.

Einstein Cross

How many galaxies do you see here? Four? Five? In fact, there is just one – the faint object in the middle of the cross. The four outer dots are all images of a quasar that lies far behind the galactic core.

This trick of the light is called gravitational lensing. In the early twentieth century Einstein predicted that gravity could bend light, as part of his theory of general relativity. This picture confirms his prediction. The galaxy's powerful gravity acts as a lens that bends and amplifies the light from the quasar behind it, producing four images of the distant object. We are literally seeing round a cosmic corner.

Galaxy NGC 7674 (top) interacting with Arp 182 (bottom)

This stunning Hubble image shows spiral galaxy NGC 7674 nearly face-on. Its central bar is made up of stars, while the long trails streaming away from the galaxy are drawn out by the gravitational pull of nearby galaxies. NGC 7674 is 400 million light-years away from Earth.

Location: Northern Hemisphere
Coordinates: Right Ascension 03h, Declination +45°

Perseus was the son of the Greek god Zeus and the mortal princess Danae. Perseus' grandfather, Acrisius, the king of Argos, was warned by an oracle that he would be killed by his grandson. When Perseus was born, King Acrisius locked the child in a wooden chest and threw it into the sea. The chest was rescued by a fisherman who raised Perseus.

Perseus had many adventures. The Gorgon Medusa had snakes for hair and her glance turned you to stone. Perseus flew to the Gorgon's den with his winged sandals, and, using his shield as a mirror, cut off Medusa's head. He later rescued the beautiful Andromeda from Poseidon's sea monster and married her. King Acrisius could not escape his fate and was accidentally killed by a discus thrown by Perseus. The constellation of Perseus is usually depicted showing him holding Medusa's head, with the bright star Algol marking her eye.

Perseus Galaxy Cluster

The Perseus Galaxy Cluster is one of the closest clusters to Earth at 235 million light-years away. However, it is difficult to see in visible light. There are 190 galaxies in the cluster, including NGC 1275. The cluster is just a small part of the Pisces-Perseus supercluster, which contains over 1,000 galaxies.

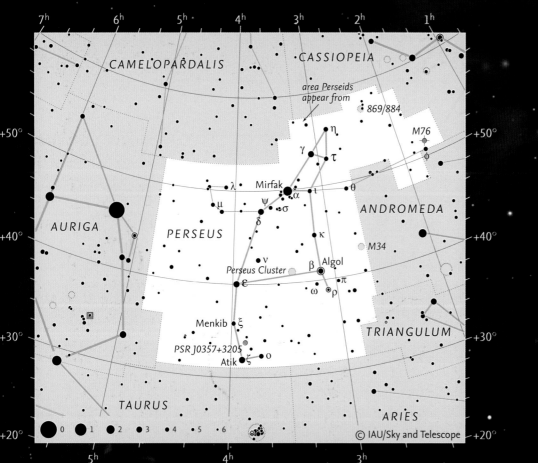

Perseus Galaxy Cluster with NGC 1275 in the centre of the image

Perseid meteor shower

Perseus A (NGC 1275)

Pulsar with a tail (PSR J0357+3205)

Perseus A Galaxy (NGC 1275) in the Perseus Galaxy Cluster

Looking a little like an intergalactic spider, this vast galaxy lies 250 million light-years away in the heart of the Perseus Cluster. This picture has been made from several images taken using different frequencies. The pink lobes of gas you can see near the middle of the galaxy were captured by radio frequencies. X-rays show the purplish shell-like formations outside them. The bright stars and background galaxies as well as the dark dust lanes were photographed by Hubble's optical camera.

Perseid meteor shower

From mid-July every year, the northern hemisphere is treated to the Perseid meteor shower. This nightly light show is caused by the dusty remains of the comet Swift-Tuttle streaming into Earth's atmosphere. During the peak period, from 9–14 August, more than 60 meteors per hour flare and die in our upper atmosphere. They have been seen for more than 2,000 years.

The puzzling pulsar

The tiny blue dot at the top right of this gassy streak is a pulsar (named PSR J0357+3205), or spinning neutron star. The streak itself is a very long X-ray tail (over 4 light-years across) that is being emitted by the pulsar. It has puzzled astronomers because it is very different from other pulsar tails. This mysterious plume is longer and more powerful than it ought to be. Astronomers hope new data from the Chandra X-ray Observatory and other telescopes will shed some more light on how this huge blue streak came to be.

CONSTELLATIONS CLOSE UP — PISCES

Location: Zodiac constellation, visible from both Hemispheres
Coordinates: Right Ascension 01h, Declination +15°

The stars in Pisces were associated with a fish (or two fish) by many ancient Middle East civilizations. The name Pisces is Latin for fish. The Greeks and Romans associated the constellation with Aphrodite and her son Eros. Aphrodite and Eros took the form of fish to escape from a monster. They tied themselves together with a string to avoid being separated. In the constellation, a bright star midway between the two fish represents the knot.

M74 – the perfect spiral

In this image from the Hubble Space Telescope, the galaxy M74 appears as an elegant spiral shell. The bright pink regions in the spiral arms are vast clouds of hydrogen gas that shine with the radiation thrown out by hot newly born stars.

M74 lies 32 million light-years away and is the dominant member of a small gathering of half a dozen galaxies. M74 is home to around 100 billion stars, making it a little smaller than our own Milky Way. Its spiral arms are so clear that it is a very popular galaxy for astronomers to study.

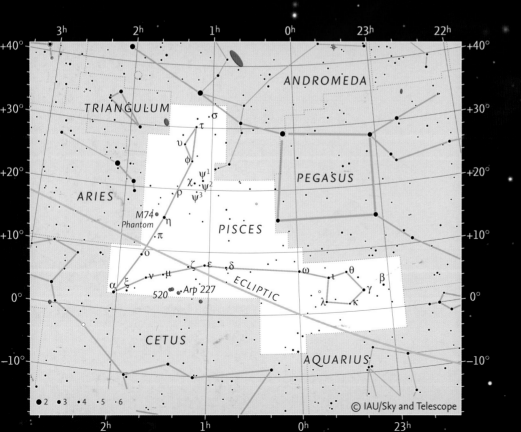

© IAU/Sky and Telescope

Phantom Galaxy (M74)

NGC 520

Arp 227 made up of NGC 474 (top) and NGC 470 (bottom)

The slow-motion smash-up

This spray of light is NGC 520, a collision of two disk galaxies that began 300 million years ago. It's a snapshot of the crash – the outer disks have joined but the inner nuclei are still separate.

Its long tail of stars stretches for 100,000 light-years. The dark brown dust lane that runs across it blocks out much of the stars beyond. Despite this, NGC 520 is one of the brightest galaxy pairs on the sky, and you can see it with a small telescope.

Ripples on a galactic pond

Swimming 100 million light-years away within Pisces is a system of galaxies known as Arp 227. This includes the galaxy at the top

(NGC 474) and the bottom (NGC 470) of this picture. NGC 470 is surrounded by wide, shell-like arcs that may have been formed by a gravitational brush with its neighbour NGC 474. The shells might also have been made when a smaller galaxy merged with a larger one. The result would have been a series of cosmic ripples, a bit like a stone being thrown into a pond.

Our image combines infrared data from NASA's Spitzer Space Telescope and ultraviolet imaging from NASA's Galaxy Evolution Explorer (GALEX) spacecraft. The different wavelengths of light reveal different features of the galaxy, giving astronomers a much more complete picture.

CONSTELLATIONS CLOSE UP — SAGITTARIUS

Location: Zodiac constellation, visible in both hemispheres
Coordinates: Right Ascension 19h, Declination -25°

Sagittarius is Latin for the archer. The constellation is often pictured as a centaur, a war-like creature with the torso of a man and the body of a horse, drawing a bow. Sagittarius represents the mythical centaur Chiron who was renowned for his gentleness.

Chiron was accidentally wounded by Hercules, who shot him with a poisoned arrow. The centaur was in agony but couldn't find the release of death because he was immortal. So he offered himself as a substitute for Prometheus. The gods had punished Prometheus for giving fire to man by chaining him to a rock. Every day an eagle ate his liver, which then grew back every night. Zeus, king of the gods, agreed. Chiron gave up his immortality and replaced Prometheus. To recognise his goodness, Zeus placed him in the stars.

The centre of the Milky Way lies near Sagittarius, so this is where our galaxy is most dense. That means there are lots of star clusters and nebulae in Sagittarius.

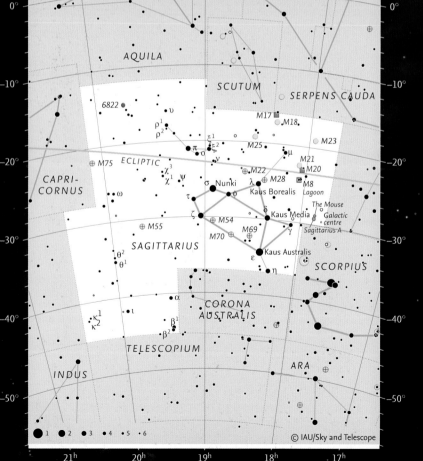

© IAU/Sky and Telescope

Lagoon Nebula (M8)

The Mouse

Sgr A West

Sgr A*

Sgr A East

Sagittarius A, X-ray

Arched Filaments

X-ray Binary
1E 1743.1-2843

Arches Cluster

Sickle

Quintuplet Cluster

Pistol Star

Sagittarius A

50 light-years

15.3 parsecs   6'35"

A view of Sagittarius A, and the galactic centre

The Lagoon Nebula (M8)
Are these tornadoes? In a way, but these interstellar 'twisters' are half a light-year long in the heart of the Lagoon Nebula 5,000 light-years away.

Hot stars in the nebula blast out radiation that strips ions away from hydrogen gas. The gas glows and throws out violent stellar winds. These space-gusts tear into cooler clouds creating weird funnels and twisted-rope structures.

The mouse that soared
G359.23-0.82 is a pulsar wind nebula that appears to have a compact snout, rounded body and a thin tail that extends for 55 light-years. So, happily, astronomers have been able to give it the much catchier name 'The Mouse'.

Pulsars throw out streams of high-energy particles that create large, magnetized clouds called pulsar wind nebulae. These are swept back as the pulsar ploughs through the interstellar gas. The Mouse is a rare young pulsar that is rocketing through space at 2 million km/h.

Sagittarius A
Deep at the heart of our galaxy a supernova and a black hole tell the story of a very stormy break up. The bright yellow and orange area in the middle of this Chandra X-ray image shows the emission of a supernova remnant, called Sagittarius A East. The white dots in the lower-right portion of the central object surround a supermassive black hole, called Sagittarius A*. Around 50,000 years ago, a star got too close to the black hole and was squished by gravity as it passed. This caused it to explode with 100 times the violence of a standard supernova, creating the scattered debris we see today

Scorpius is Latin for scorpion. Many Greek myths about Scorpio also featured Orion, the hunter. In one, Orion boasted to goddess Artemis and her mother, Leto, that he would kill every animal on the earth. Artemis was also a keen hunter, but this time she offered to protect all creatures. Artemis and Leto then sent a scorpion to fight Orion.

example to mortals to restrain their pride.

Antares and M4

Antares is the brightest star in the constellation of Scorpius and one of the brighter stars in the night sky. Located 550 light-years away, it is a huge red supergiant about 850 times the diameter of our own Sun, 15 times more massive, and 10,000 times brighter. Antares is seen on the left surrounded by a yellowish nebula of

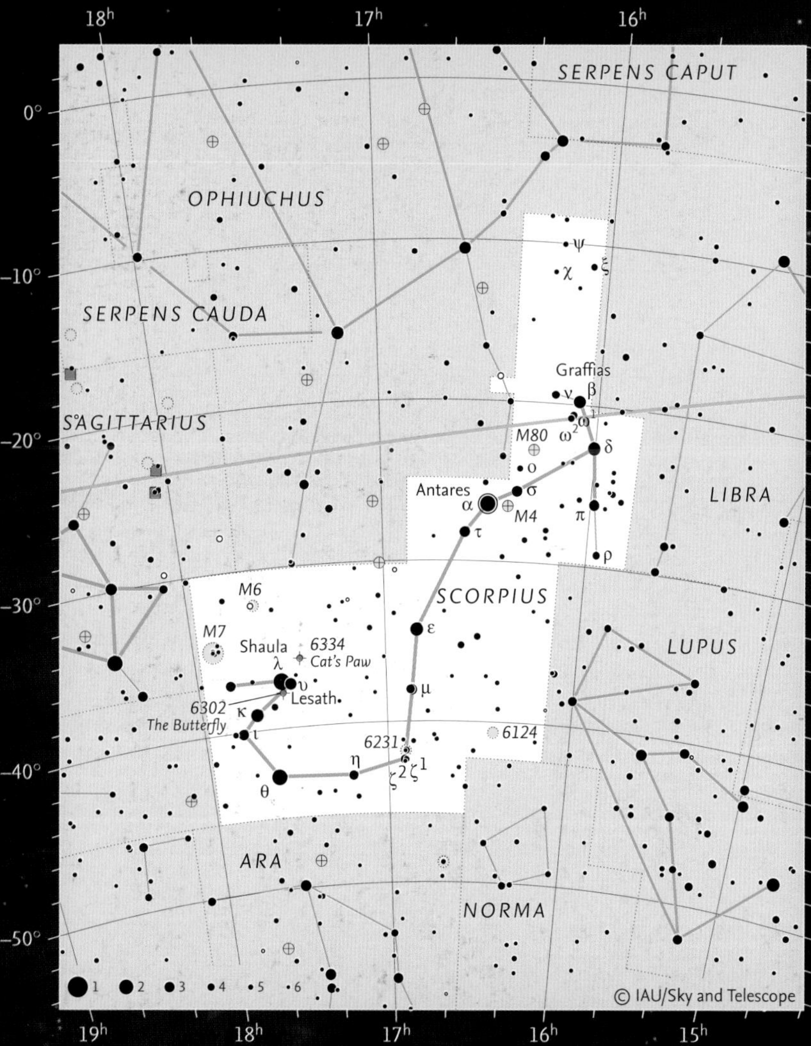

© IAU/Sky and Telescope

Antares/M4 region

Cat's Paw Nebula (NGC 6334)

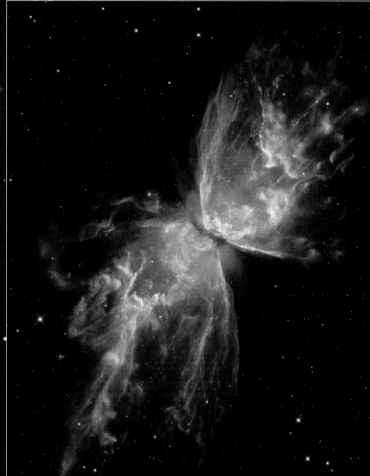
The Butterfly Nebula (NGC 6302)

gas that it has expelled. Radiation from a nearby blue star helps the nebular gas glow. At the bottom of the image is the globular star cloud M4, which lies behind Antares. The bright star on the far right is Al Niyat, which is named after the Arabic for 'the arteries [of the scorpion]'.

The Butterfly Nebula NGC 6302

It looks like a delicate butterfly, but this nebula is a super-violent cosmic firestorm. The nebula represents the death-throes of a hidden star, which at 250,000 °C is one of the hottest known stars in our galaxy. The dainty 'wings' are actually churning cauldrons of superhot gas streaming across space at a million km/hour. They would speed from the Earth to the Moon in 24 minutes! The butterfly itself stretches for more than two light-years, around ha the distance from the Sun to the nearest star, Proxima Centauri.

Cat's Paw Nebula

The padded shapes of this well-named emission nebula are reddish puffy clouds of glowing gas. The Cat's Paw Nebula (NGC 6334) looks to lie near the centre of the Milky Way, but is actually close to Earth, at just 5,500 light-years away.

Reaching across the cosmos for 50 light-years, the Cat's Paw is one of the most active star formation regions in our galaxy. It is home to tens of thousands of stars, most of them huge brilliant blue stars that have formed in the last few million years.

on: Zodiac constellation, visible in both Hemispheres

inates: Right Ascension 04h, Declination +15°

, the bull, is among the very oldest of the recognised llations. Some researchers claim that Taurus is represented ve painting at Lascaux in France, which dates to 15,000 BC.

ancient Mediterranean and Middle East civilizations were at eights, the Sun entered Taurus in the spring, making it an cant symbol for these herding and agricultural societies.

eiades is a cluster of seven bright stars in Taurus that acted ther forecasters. They were only visible when the skies were

very clear, so ancient seamen would start their sea voyages when they appeared. The name 'Seven Sisters' has been used for the Pleiades in different cultures stretching from Australia to North America and Siberia. This suggests the name came from one single ancient origin.

The brightest member of this constellation is Aldebaran, an orange-hued giant star. Its name comes from the Arabic for 'the follower' of the Pleiades. Aldebaran forms the bloodshot eye in th A-shaped face of Taurus. The bull seems to glare menacingly at the hunter Orion, who lies just to the southwest.

You can best see Taurus in December and January.

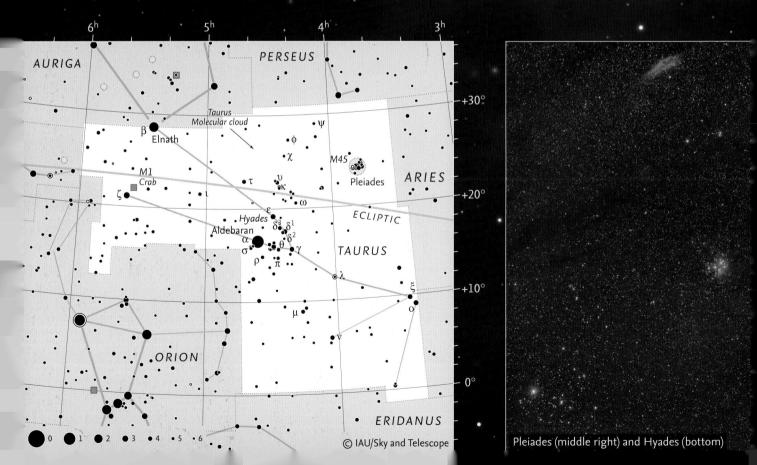

© IAU/Sky and Telescope

Pleiades (middle right) and Hyades (bottom)

The Pleiades (M45)

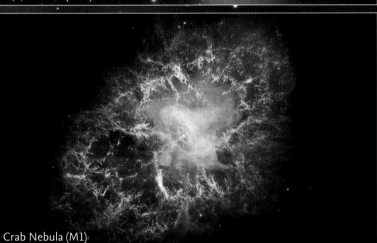

Crab Nebula (M1)

Taurus Molecular Cloud

The Pleiades (M45)

This cluster of brilliant stars is known as the Pleiades, or Seven Sisters. Named after the seven daughters of Atlas and Pleione in Greek mythology. The Pleiades is an open cluster of mostly hot blue and very luminous stars that have formed within the last 100 million years. It is one of the nearest star clusters to Earth and the easiest one to spot with the naked eye. The Pleiades will stay together for around 250 million years before being torn apart by the gravity of nearby gas and dust clouds.

Crab Nebula (M1)

The Crab Nebula is a supernova remnant of an explosion seen from Earth on 4 July 1054. The supernova was so bright it was visible during the day. The star's spectacular death was recorded by Chinese historians and Native Americans in New Mexico painted its image on a canyon wall. Today, almost a thousand years later, we can see the super dense neutron star left behind by the explosion spewing out a blizzard of high-energy particles. This cosmic generator is producing the energy of 100,000 Suns. This picture of the Crab Nebula is a composite of X-ray, optical and infrared images.

Taurus Molecular Cloud

This smoky wisp is a filament of cosmic dust more than ten light-years long, which is just part of the Taurus Molecular Cloud. Baby stars snuggle in the Cloud's folds, which are dense plumes of gas ready to collapse and form yet more stars.

CONSTELLATIONS CLOSE UP — URSA MAJOR

Location: Northern Hemisphere
Coordinates: Right Ascension 11h, Declination +50°

Ursa Major means 'Great Bear' in Latin. The Greeks imagined the stars of this constellation forming the shape of a bear walking about on its clawed feet. With its companion Ursa Minor it was said to be the prey of Boötes and his hunting dogs.

You can easily spot its most famous feature, the seven-star shape known as the Plough. This looks like a saucepan with a handle, or a ladle, hence its other name – the Big Dipper. You can see it best in April.

The two stars at the front of the Dipper, Merak and Dubhe, point almost directly north to Polaris, which has been a useful marker for generations of travellers.

The dynamic galactic duo

M81 is a great spiral galaxy, similar in size and brightness to our Milky Way. The stars in its spiral arms have formed within the last 100 million years. But in the nucleus of M81, the stars are 10 billion years old, and are near the end of their lives. They produce their energy by burning helium into carbon. At the centre of M81 is a supermassive black hole that is about 70 million times more massive than the Sun.

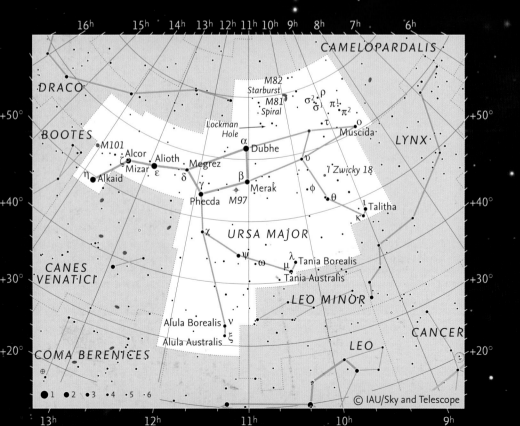

© IAU/Sky and Telescope

The Great Bear

The Lockman Hole

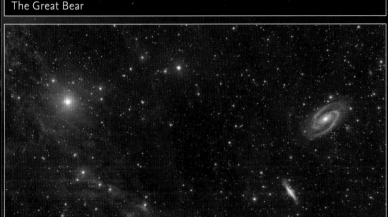
M81 and M82, on right hand side of image

I Zwicky 18

Starburst central

M82 is a starburst galaxy, where stars are forming hundreds of times faster than in a normal galaxy. Star formation is so violent that gas and dust are being jettisoned perpendicular to its disk. Supernova explosions are shooting out bubbles of hot gas millions of light years long.

The frenzy of formation was triggered by a close encounter with nearby galaxy M81 millions of years ago. The massive shock wave sent star formation into overdrive. New images from the Chandra Observatory suggest that the galaxy also contains at least eight black holes.

The Lockman Hole

A starless patch of the sky might not seem that interesting, but a part of the sky called the 'Lockman Hole' has been very useful to astronomers. This area contains almost no objects within our Milky Way galaxy, so astronomers can look through it, deep into space, to study galaxies in the distant Universe. All of the little dots in our picture are distant galaxies, rather than stars. The pattern of their light is called the cosmic infrared background. By studying it, astronomers measured how much dark matter it takes to create a galaxy bursting with young stars.

The baby that suddenly grew up

Dwarf galaxy I Zwicky 18 was once thought to be a mere toddler, containing young bluish-white stars only found in the early Universe. But images from the Hubble Space Telescope showed that it was really a fully-fledged adult galaxy. Hubble found reddish stars that may be 10 billion years old, suggesting that I Zwicky 18 formed at the same time as most other galaxies.

CONSTELLATIONS CLOSE UP — URSA MINOR

Location: Northern Hemisphere
Coordinates: Right Ascension 15h, Declination +75°

Ursa Minor means 'Smaller Bear' in Latin. Like the Great Bear, its tail may be seen as the handle of a ladle, so it's sometimes called the Little Dipper. The two heavenly bears were said to be hunted by Boötes and his dogs.

The brightest star in the constellation is Polaris. This is a unique star in the heavens. It is known as the 'north star' because it is very close to the north celestial pole, the spot in the heavens above Earth's own north pole. Its position there means that it stands motionless in the sky while all the other stars appear to us to spin round it. You can see from our time-lapse picture how the light-trails of other stars form circular paths centred on Polaris.

For centuries, Polaris has worked as a celestial sat-nav. All you have to do is find the star and you know where north is. You can then determine south, east and west and generally work out which way you're going! One of its ancient names is the 'lodestar' meaning guiding star.

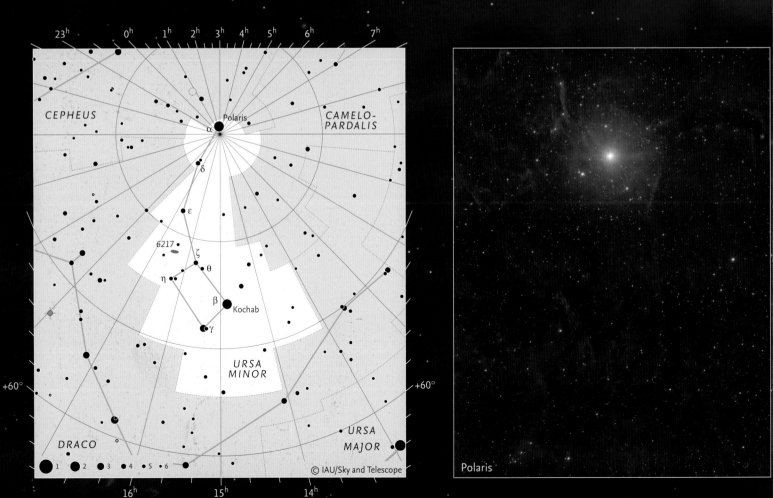

© IAU/Sky and Telescope

Polaris

Time-lapse star trail around Polaris

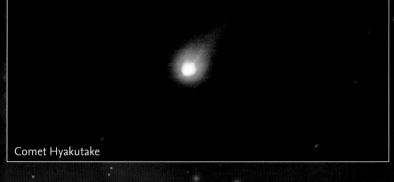
Comet Hyakutake

Barred spiral galaxy NGC 6217

But it won't always be in this helpful position. By the end of the twenty-first century, the celestial pole will move away from Polaris, and by the forty-first century our descendants will be looking to the star Gamma Cephei to find north.

In our picture, Polaris is at the centre of a delicate, complex web of dust. These molecular clouds are sometimes known as 'galactic cirrus', after the high wispy clouds in Earth's atmosphere.

Barred spiral galaxy NGC 6217

Many spiral galaxies have bars across their centres, including our own Milky Way galaxy. Spiral galaxy NGC 6217 has a very large, clearly defined bar, as this spectacular image taken by the Advanced

You can see the dark, spindly dust lanes, clusters of bright blue young stars and red emission nebulae of glowing hydrogen gas. The shining nucleus in the central bar is probably home to a supermassive black hole. Light takes about 60 million years to reach us from NGC 6217, and the galaxy is 30,000 light-years across.

Comet Hyakutake

This comet was discovered on 31 January 1996 by an amateur astronomer using a pair of binoculars! Two months later it came as close to Earth as any comet has in 200 years. Hyakutake became very bright in the night sky and was seen world-wide. It will next visit us in 70,000 years!

ion: Zodiac constellation, visible in both Hemispheres

linates: Right Ascension 13h, Declination 0°

is Latin for virgin. The constellation was linked to
e deities in many early civilizations. The constellation rose
d-August for the ancient Greeks and Babylonians, making
mportant marker for the end of summer and the ripening
harvest.

is the second largest constellation after Hydra and is quite
o spot using its bright star Spica. Follow the curve of the Big
r to Arcturus in Boötes and keep on in a similar curve to find
'Follow the arc to Arcturus and speed on to Spica.'

Gravitational lens and galaxy cluster, MACSJ 1206

Every bright object in this picture is a galaxy with 100 billion stars.
This cluster of galaxies is so immense that it bends space itself. It
might look random at first, but look closely and you will see that
there is a clear circular pattern. That's because the cluster acts as a
lens, warping the light of further galaxies into multiple images and
arc-like smears.

Eruption of a galactic super-volcano

M87 is a supergiant elliptical galaxy with a huge black hole at
its core that is spurting out massive jets of energetic particles.
The way these jets interact with hot gas is similar to the eruption
of a volcano on Earth.

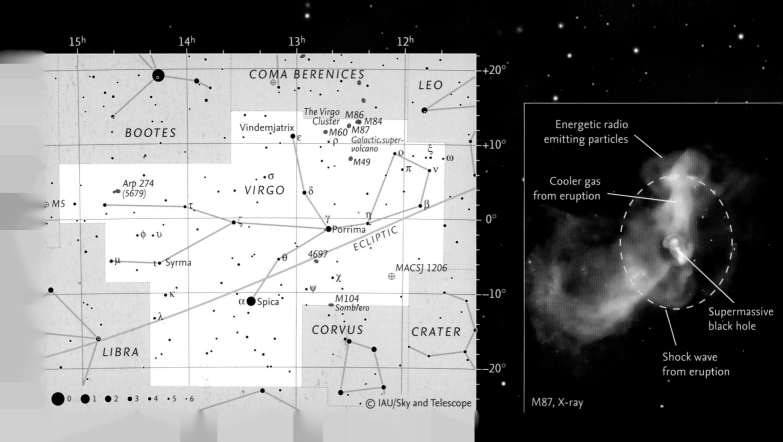

© IAU/Sky and Telescope

Energetic radio
emitting particles

Cooler gas
from eruption

Supermassive
black hole

Shock wave
from eruption

M87, X-ray

MACSJ 1206

Sombrero Galaxy (M104)

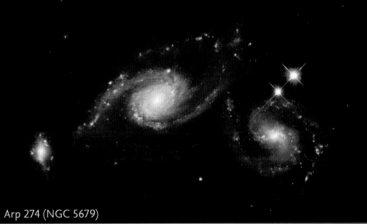

Arp 274 (NGC 5679)

When the Eyjafjallajökull volcano in Iceland erupted in 2010, pockets of hot gas blasted through the surface of the lava, generating shock waves in the grey smoke of the volcano. This hot gas then rose and dragged dark ash with it.

So the energetic jets from the black hole rise through the galactic gas, creating shock waves and lifting up the coolest gas clouds in their wake.

Hats off to M104

It's easy to see why astronomers call this stunning galaxy the 'Sombrero'. With its bulging nucleus of mature stars, and its swooping disk composed of stars, gas, and dust, this has to be one of the most beautiful of all galaxies. You can also see dark lanes in the dust, where future stars will be born.

Many of the small firefly-like bright spots in the Sombrero's halo are globular clusters, similar to those found in our own galaxy.

Three swirling star factories

Arp 274 (or NGC 5679) is a system of three galaxies that seem to be overlapping, but are actually spread out in space.

The two larger spiral galaxies are giving birth to baby stars at a phenomenal rate. You can see this in the bright blue knots strung along the galactic arms. These are groups of young stars. Around them are pink nebulae that are lit up by the young stars. Older yellow stars cluster in the central galactic bulges. The more compact galaxy on the left doesn't have spiral arms but it is clearly also creating new suns.

OBSERVING THE PLANETS

So how do you start actually looking at the planets? Well, you can see Mercury, Venus, Jupiter and Saturn with the naked eye. But you can observe so much more with a good pair of binoculars or a telescope.

You will also need to arm yourself with some information about where to look and what to look for. You can find these details for Mercury, Venus, Mars, Jupiter and Saturn in the following pages.

There are also lots of websites that have the latest information about what's hot in the heavens.

Check out these sites:
www.astronomytoday.com/skyguide.html
astro-observer.com/index.html

Should I buy a telescope?
There are an incredible variety of telescopes on the market. Even experts find it hard to choose the right instrument, so it can be a minefield for beginners. One firm rule is that you get what you pay for. If a scope is cheap, it is likely to be of lower quality. Be wary of buying without trying, particularly second hand. The scope could be damaged.

We hope you will enjoy stargazing, but it may not be the hobby for you. So before you spend lots of money on a scope, why not start with a good pair of 10x50 binoculars? Many professional astronomers did just that when they were young. You can also contact your local astronomy society – some of its members may be happy to give you advice and let you try out their telescopes.

Skywatching tips

Find the darkest sky you can. Modern streetlights fill city skies with an orange glow that interferes with stargazing, as you can see from our picture of Phoenix. You can still see the planets, but you will get a better view in a dark open space. Going to a public park can help, and if you live near the countryside that is even better.

You can also find out if there is a dark sky park near you. These are locations that are officially recognised for the exceptional beauty of their night skies. To find dark sky parks, visit www.darksky.org/night-sky-conservation/national-park-service

Learn your way around the night sky. You can get computer-controlled telescopes that 'Go To' any location you program. But you will learn more about the stars and ultimately get greater enjoyment by using a star map to navigate around the heavens. There are many websites and apps that help you learn the layout of the skies.

Get warm! Clear skies are ideal for astronomy, but they also mean it will be cold! Since you will be spending most of your time sitting still, it's vital that you keep warm. Wear lots of layers, a hat and gloves and take a thermos of hot tea or soup.

Acclimatize your telescope. Bring your scope out at least 30–60 minutes before you start observing. This will cool it down to the ambient temperature, stopping air currents inside your scope from making the image go wobbly.

OBSERVING MERCURY

Mercury is the planet closest to the Sun, and also the smallest planet. It is known as an inferior planet because its distance from the Sun is less than the Earth's. Because it orbits between our planet and the Sun, it has phases. In other words, its shape appears to change over time, like the Moon. Really this is because we see different amounts of sunlight falling on it as it moves relative to the Sun.

Where to find it

You can see Mercury with the naked eye. It appears in the west after sunset and in the east before dawn. The planet always stays close to the Sun. You get the best views of the planet in spring evenings and autumn mornings, when the planet appears higher above the horizon.

Because Mercury is so close to the Sun it can be quite tricky to spot. You must always be exceptionally careful when looking for Mercury that you don't accidentally point your binoculars or telescope at the Sun. We are not being dramatic when we say that you can be instantly and permanently blinded by looking at the Sun through a magnifying device. Please be very, very careful.

What to look for

Mercury zooms round the Sun every 88 days, and occasionally it makes a transit. This is when it moves across the face of the Sun, appearing as a small black disk crawling in front of the very bright sphere. In our time-lapse image you can see the path of the planet – and how small it is compared with our star!

The planet has mountains, ridges, valleys and thousands of craters, making it look a little like the Moon. The Mariner 10 probe flew past Mercury in 1974, and NASA's MESSENGER spacecraft became the first ever to orbit the planet in 2011. MESSENGER discovered water ice at Mercury's north pole and craters filled with lava, showing the planet was once volcanic. In its first year-long mission it sent back nearly 100,000 photographs.

Time-lapse image of Mercury in transit across the face of the Sun above Venus with the moon at the bottom beside the silhouette of the telescope's housing. So in this image you can see all the major celestial objects that pass between Earth and the Sun.

Image from the VLT observatory in Chile

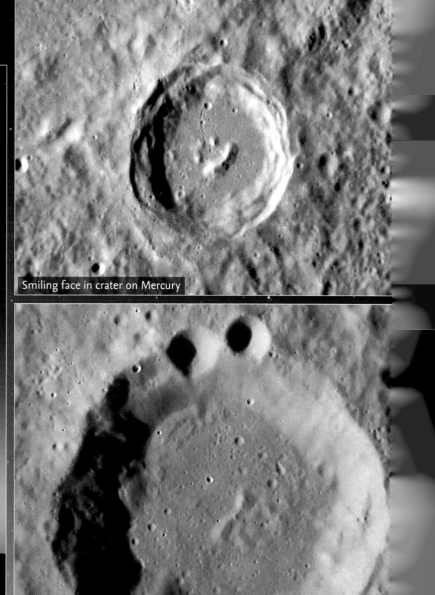

Smiling face in crater on Mercury

Cookie monster crater on Mercury

dawn. You can't see it in the darkest part of the night, and it doesn't rise very high over the horizon.

Where to find it

When Venus is on one side of the Sun, it trails the Sun in the sky and appears just after sundown. This is when Venus is the Evening Star. Because Venus orbits the Sun faster than Earth it 'overtakes' us every 584 days. At this point it changes from being the 'Evening Star' to the 'Morning Star'. Venus is now on the other side of the Sun, leading it across the sky. Venus will rise a few hours before the Sun in the morning. When the Sun rises and the sky lightens, Venus fades away.

What to look for

Venus is like Mercury in that it orbits nearer to the Sun than Earth. So we can see that it has distinct phases. When it is nearest to us it shines as a slender crescent. It then waxes gibbous as it moves further from us. Venus also makes transits across the Sun like Mercury. However, these only occur rarely – the next one will be in the year 2117!

This is pretty much where the similarities between the two inferior planets end. Mercury is small and rocky with no atmosphere.

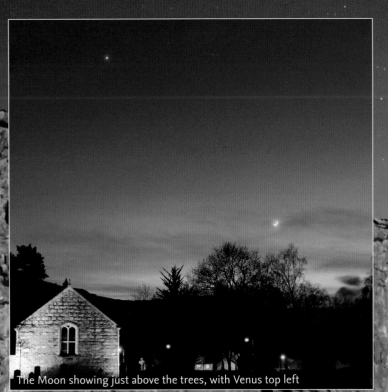

The Moon showing just above the trees, with Venus top left

Transit of Venus across the Sun

Venus is nearly as big as the Earth and has a dense atmosphere. This thick atmosphere is a problem for astronomers. It's a toxic stew of carbon dioxide and sulphuric acid clouds that's impossible to see through from Earth. Despite being our nearest planetary neighbour and coming as close as 38 million km, all we ever see of Venus is the top of a thick cloud layer, which is reflecting sunlight.

It wasn't until spacecraft visited Venus that we saw what it was really like – a world of towering mountains, rolling plains, and majestic canyons. More than twenty missions have successfully reached the planet. The Magellan probe saw through the planet's clouds by using radar. It mapped 99 per cent of the planet's surface in the early 1990s. Other craft have landed and sent back pictures for a short time before being crushed by the planet's immense atmospheric pressure.

Three impact craters on the surface of Venus have become known as the 'Crater Farm'

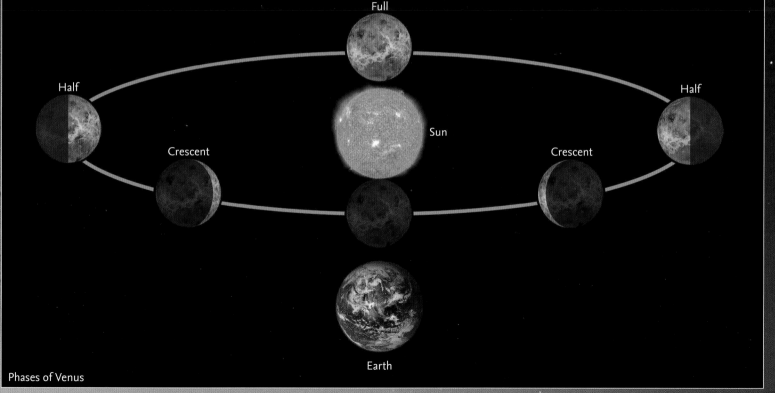

Full

Half

Crescent

Sun

Crescent

Half

Earth

Phases of Venus

Mars is the next planet outwards from the Earth, but it is a small world at 6,800 km in diameter. It has a wealth of interesting surface features, with dark patches like continents among an ocean of pink deserts, polar ice caps, clouds, and dust storms.

Where to find it

Mars moves around the sky more than the inferior planets, and you can check an astronomy website to find out where it is on a particular date (try astronomycentral.co.uk/planets-to-see-in-the-sky-tonight).

What to look for

You can see the red hue of Mars through binoculars but you will need a telescope to pick out any detail. One fun way to see more of the planet is observe it several times during a 30-60 minute period and sketch what you see each time. You will probably see different details each time you look, so if you add these to the same sketch you will build up a nice picture of the planet.

The most obvious feature of Mars is its colour, a bright, peach-like hue that is actually a vast global desert. Mars also has bright ice caps at its poles. It is interesting to watch as they shrink in the Martian spring and return in winter. The northern ice cap never completely vanishes, but the southern one does.

It's also a good idea to look at Mars often, and at different times of night. This is because Mars rotates every 24 hours and 37 minutes, a period not far off our own day. So that means you see virtually the same view of the planet at the same time each night. By varying the times you look at it you can see the different sides of the planet.

Mars has been well studied by astronomers and many spacecraft have visited the planet. In the Hubble Space Telescope image of Mars you can clearly see the planet's south polar ice cap.

The southern ice cap on Mars shrinks and then vanishes each year

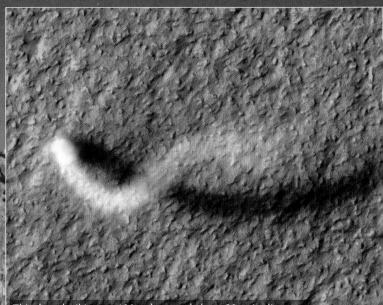

This dust devil is over 600 m long and about 30 m in diameter

Let's twist again

Like on Earth, winds on Mars get their power from the Sun heating the planet's surface. So you can see many familiar wind features in the Martian atmosphere – like this dust devil. You can see this twister casting a serpentine shadow; this was caused by a westerly breeze partway up the plume. The plume reached 800 m above the surface and was 30 m in diameter.

View from the surface

NASA's Curiosity rover is the latest spacecraft to send back amazing images of the red planet's surface. This panoramic scene, snapped in 2012, could be from a desert in Arizona. It shows the view taken while the rover was working at a site called 'Rocknest'. The rover was scoping out this area to select a good site for drilling into the Martian rocks.

The Martian landscape at 'Rocknest' photographed by the Curiosity rover with its Mastcam

Watch out, planet reversing

Sometimes Mars appears to stop in its orbit, go backwards, then change direction again to form a little loop before continuing on its way. But the planet isn't really dancing, this effect is called retrograde motion and is an effect caused by Earth 'overtaking' planets that orbit farther from the Sun.

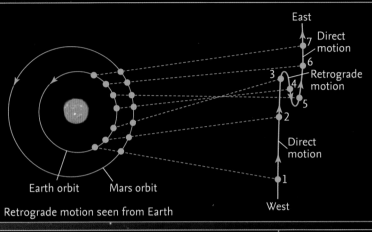

Retrograde motion seen from Earth

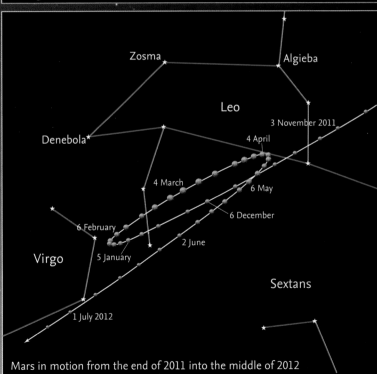

Mars in motion from the end of 2011 into the middle of 2012

Jupiter is a very rewarding planet to watch. It's the third brightest object in the night sky, after the Moon and Venus. Even when it's far away from Earth, it's such a large planet that you will still be able to pick out lots of detail.

Where to see it and what to look for
Like Mars, you are best checking an astronomy website for Jupiter's latest location (try astronomycentral.co.uk/planets-to-see-in-the-sky-tonight). When you find it, you will notice it spins round completely every 10 hours, so even during one night you can see different areas of the planet. As Jupiter is a gas giant the features we see are moving parts of its atmosphere, rather than a solid surface, so it won't look exactly the same two nights in a row

What an atmosphere
There are many distinct bands of clouds, the two clearest of which lie on either side of the planet's equator, and so are known as the equatorial belts. Storms often appear in these belts, looking like beads on a string.

The Moon with Venus below it and Jupiter midway up on the left side

The Great Red Spot with 'Red Spot Jr' below and 'Baby Red Spot' to the left

Jupiter's most famous feature is the Great Red Spot. This fearsome red storm is bigger than Earth. It varies in intensity and stays in view for only a few hours, giving you a great sensation of Jupiter's swift rotation.

There are other spots too, constantly developing, merging and disappearing. The 'Red Spot Jr' appeared in 2006 and has had two narrow escapes, passing close by its big brother unharmed. But the 'Baby Red Spot' that appeared two years later was caught up in the larger vortex in 2008 and destroyed.

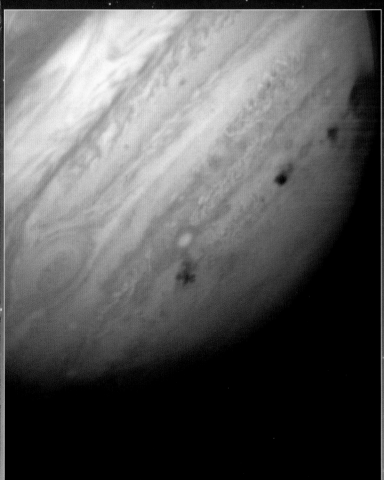

Dark patches on Jupiter's surface were caused by the impact of fragments of Comet Shoemaker-Levy 9

Moon spotting

At least 67 moons orbit Jupiter, but most of these are too faint to see. But you can spot the four Galilean moons – Ganymede, Io, Europa and Callisto – using just a pair of binoculars and a steady hand. Your camera phone may also be able to pick up the planet as a bright dot. Some impressive pictures of Jupiter have been taken simply by securing a camera phone to the eyepiece of a telescope.

Ganymede is larger than the planet Mercury, while Io is larger than our own Moon, and Europa and Callisto also shine brightly. They are often up to something interesting, and you can watch them make a transit across the face of the planet, pass behind it and reappear or fall into eclipse when they move into its shadow. They sometimes cast their own tiny shadows on Jupiter's face. It is fun to watch them change positions from night to night.

Crash, bang, wallop

Poor Jupiter is a bit like an elephant wandering about a golf driving range. It's so big and so close to the asteroid belt that it is going to get hit by something fairly often. In 1994 it was the turn of comet Shoemaker-Levy 9. This was torn into fragments by Jupiter's strong gravity before it smashed into the gas giant. In our image you can see the huge dirty bruises left by the impact.

The most beautiful sight in a telescope has to be Saturn. The planet's golden globe is surrounded by its crowning glory – crystal white rings set against the blackness of space.

Where to see it and what to look for
Like Jupiter, you are best checking an astronomy website for Saturn's latest location (try astronomycentral.co.uk/planets-to-see-in-the-sky-tonight). You will see that the planet's body has belts like Jupiter, but they are much fainter. Sometimes it also has storms. If you watch Saturn regularly you will quickly learn to spot these when they appear.

Running rings around!
Saturn isn't the only planet to have rings, but it does have the brightest. This is because they are made of water ice particles. These stay shiny because they are often bashing into each other, splitting and reforming to show new faces that reflect sunlight.

Saturn's rings are most easily seen when they are at an angle

The rings are very clear when the Sun is behind Saturn

There are many distinct rings, named with letters in order of their discovery (A-G). Galileo glimpsed them in 1610, but couldn't tell what they were. He described Saturn as having 'ears'! Christiaan Huygens was the first person to describe them as a disk in 1655.

There are also many distinct black gaps between the rings. Many of these are created by Saturn's moons. As the icy lumps pass the moons they are tugged a little by gravity. The moons 'sweep clean' neat bands in the rings. Your view of the rings will depend on how much the planet is tilted towards us. When the rings are face-on, look for the dark Cassini Division between the A and B rings. This tiny black gap is actually as wide as the Atlantic Ocean!

When the rings are edge-on you won't be able to see them, but you will notice how squashed Saturn is. The planet's diameter at its poles is 10 per cent less than at its equator. Imagine holding a squash ball between your thumb and forefinger and squeezing slightly. Saturn is said to be 'oblate'. This is caused by Saturn's very low density and very high rotation speed. It spins fully round every 10 hours 38 minutes.

Saturn has sixty-two moons, but you will probably be able to see up to eight; the rest are too faint. The easiest moon to see is Titan, the biggest and brightest. If you catch it transiting Saturn, you may be able to see the orange glow of its smoggy atmosphere. Iapetus is a two-toned moon that has one side bright as snow and the other as dark as coal.

The Voyager I probe visited Saturn in 1980, sending back the first detailed pictures. The Cassini-Huygens spacecraft has returned amazing pictures of Saturn and its moons since 2004. It has observed water geysers on the moon Enceladus and hydrocarbon seas on Titan. The Huygens part of the probe landed on Titan in 2005 and returned a wealth of images.

Four of Saturn's moons caught in transit, this is a relatively rare occurrence

Titan's surface appears to have polar seas, but these are not water, this is Ligeia Mare, a hydrocarbon lake that is bigger than Lake Superior

MAN AND SPACE

Sputnik 1 had four radio antennae to broadcast radio pulses.

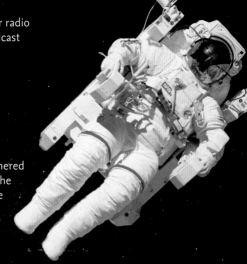

Bruce McCandless II makes the first untethered spacewalk as part of the 1984 *Challenger* Space Shuttle mission

1957 – 1969

1957 The first artificial satellite, Sputnik 1, launched by the Soviet Union

1959 First photograph of Earth from orbit from the Explorer 6 satellite

1961 First man in space, the cosmonaut Yuri Gagarin

1963 First woman in space, the cosmonaut Valentina Tershkova

1965 First spacewalk by cosmonaut Alexey Leonov

1966 First spacecraft to land on the Moon, Luna 9

1969 Man lands on the Moon, Apollo 11

1970 – 1979

1970 First lunar rover, the unmanned Lunokhod 1

1971 First manned space station, Salyut 1

1972 Last manned moon landing, Apollo 17

1975 First images of Venus from Mariner 9

1976 First images from the surface of Mars from the Viking Lander

1977 Voyager spacecraft launched

1979 First images of Jupiter from Voyager 1, and of Saturn from Pioneer 11

1980 – 1989

1980 Voyager 1 passes Saturn

1981 First Space Shuttle orbital test flight, *Colombia*

1986 Space Shuttle *Challenger* disaster, 73 seconds into the flight

1986 Voyager 2 passes Uranus

1986 First long-term research space station stay on Mir

1989 Voyager 2 passes Neptune

The Lunar Module from Apollo 11 photographed on the Moon in 1969

Mars Curiosity rover took 55 images when it was at the base of Gale Crater in October 2012, to make up this self-portrait mosaic

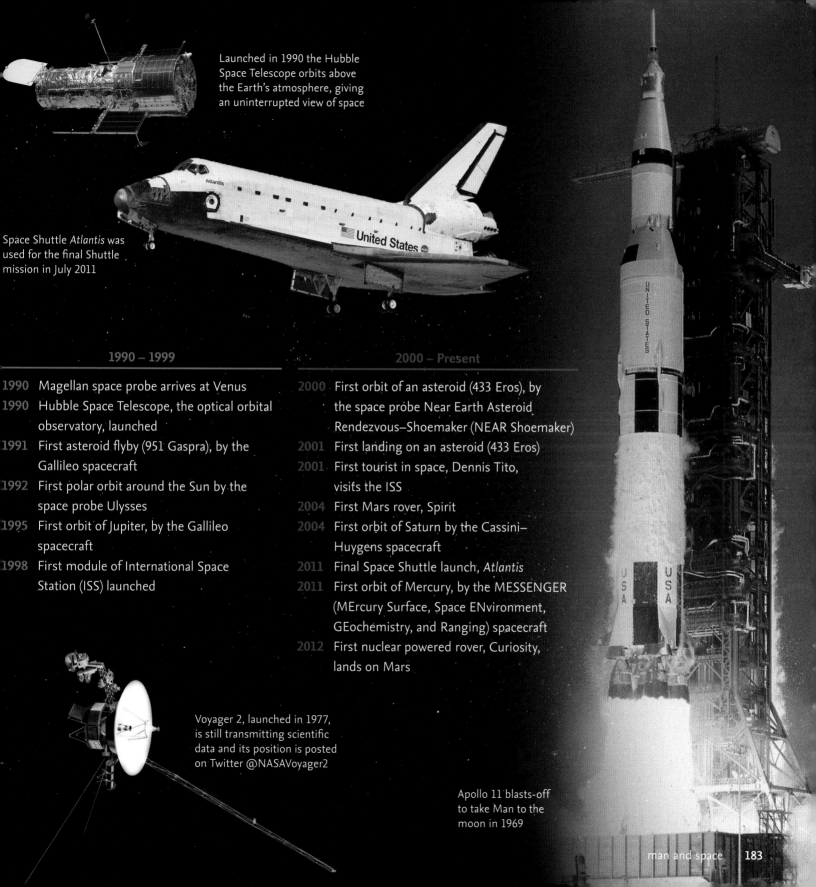

Launched in 1990 the Hubble Space Telescope orbits above the Earth's atmosphere, giving an uninterrupted view of space

Space Shuttle *Atlantis* was used for the final Shuttle mission in July 2011

1990 – 1999

1990   Magellan space probe arrives at Venus

1990   Hubble Space Telescope, the optical orbital observatory, launched

1991   First asteroid flyby (951 Gaspra), by the Gallileo spacecraft

1992   First polar orbit around the Sun by the space probe Ulysses

1995   First orbit of Jupiter, by the Gallileo spacecraft

1998   First module of International Space Station (ISS) launched

2000 – Present

2000   First orbit of an asteroid (433 Eros), by the space probe Near Earth Asteroid Rendezvous–Shoemaker (NEAR Shoemaker)

2001   First landing on an asteroid (433 Eros)

2001   First tourist in space, Dennis Tito, visits the ISS

2004   First Mars rover, Spirit

2004   First orbit of Saturn by the Cassini–Huygens spacecraft

2011   Final Space Shuttle launch, *Atlantis*

2011   First orbit of Mercury, by the MESSENGER (MErcury Surface, Space ENvironment, GEochemistry, and Ranging) spacecraft

2012   First nuclear powered rover, Curiosity, lands on Mars

Voyager 2, launched in 1977, is still transmitting scientific data and its position is posted on Twitter @NASAVoyager2

Apollo 11 blasts-off to take Man to the moon in 1969

INTERNATIONAL SPACE STATION

The International Space Station (ISS) was designed to be a laboratory, observatory and factory in space, which could act as a staging base for possible future missions to the Moon, Mars and asteroids. There have been over 130 launches to the space station since the launch of the first module, Zarya, in 1998. The space station is visible with the naked eye just before sunrise or after sunset. It moves very slowly and appears as a bright white dot. It crosses the sky in two to five minutes. It is the brightest man-made object in the sky and orbits 386 km above Earth..

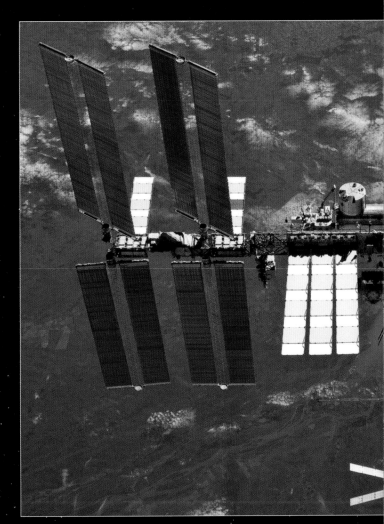

The International Space Station from the space shuttle in March 2011

International Space Station Facts

Launch Date: 1998
Module Length: 51 metres
Mass: 391 tonnes
Habitable Volume: 388 cubic metres
Power Generation: 84 kilowatts
Average speed: 32 410 km/h
Size of crew: 6

Astronaut Dave Wolf outside the ISS on his second space walk in July 2009

The International Space Station spans the area of a U.S. football field, including the end zone. It is the size of a conventional five-bedroom house, and has two bathrooms, a gym and a 360-degree bay window. The station's first resident crew, Expedition 1, marked the beginning of a permanent international human presence in space, arriving at the station in a Russian Soyuz capsule in November 2000. Currently, station crews stay on orbit for six months at a time. In November 2012, NASA and the Russian Federal Space Agency (Roscosmos) announced an agreement to send two crew-members to the space station on a one-year mission that they hope to start in the Spring of 2015.

Crew members communicate with the public via Twitter on @NASA_Astronauts and they also post some photographs of places on Earth that the space station has passed.

Images of Earth from the International Space Station

Dubai, United Arab Emirates

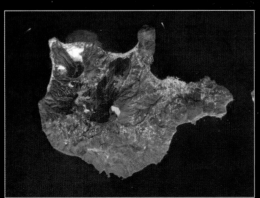

The island of Lipari, just north of Sicily in the Mediterranean Sea

Sarychev Volcano on Matua Island in the Kuril Island chain.

GLOSSARY

Altitude One of the two coordinates that can be used to pinpoint a star in the sky, the other being azimuth. The altitude is how far above or below the horizon the object lies. It is measured in degrees.

Anti-clockwise The direction that the majority of objects in our Solar System orbit the Sun, as seen from above the Sun's north pole. The Sun, Earth and most of the planets also spin on their axes in this direction.

Asterism A group of stars that form a pattern.

Asteroid A small rocky object in space. There are thousands in a belt between Mars and Jupiter. Also known as a minor planet.

Astronomer Someone who studies space, including the planets, stars, galaxies and comets.

Atmosphere The layer of gases surrounding the surface of a planet, moon or star.

Aurora A light display at the Earth's poles, known as the Aurora Borealis in the north and Aurora Australis in the south. Aurorae are caused by charged particles from the Sun entering Earth's upper atmosphere.

Azimuth One of the two coordinates that can be used to pinpoint a star in the sky, the other being altitude. The azimuth is how far round the horizon the object lies, relative to north. It is measured in degrees.

Binary star Two stars bound together by gravity. They orbit each other, taking anywhere from half an hour to millions of years to make a circuit.

Billion 1,000,000,000 or a thousand million.

Black hole An area of space round a collapsed star that has such a strong gravitational pull that nothing, not even light, can escape it.

Celestial sphere An imaginary sphere surrounding the Earth. It is used by astronomers to mark the positions of space objects.

Chandra Observatory NASA's Chandra X-ray Observatory is a space telescope launched in 1999. It detects X-rays from very hot areas like exploded stars, galaxy clusters, and matter around black holes.

Circumference The distance round the outside of a sphere or circle.

Cluster (of galaxies) A group of galaxies bound together by gravity. Clusters can contain many thousand individual galaxies.

Comet A small lump of ice and dust that orbits the Sun (a few orbit Jupiter). When comets approach the Sun, some of their material boils off into a long tail.

Conjunction When a planet appears close to another planet or star in the sky it is said to be in conjunction.

Constellation A group of stars as seen on the celestial sphere. They are often named after mythical characters and creatures.

Core The centre of a space object, such as a planet, moon or star.

Corona The outer part of the Sun's atmosphere, best seen during a solar eclipse. It is faint but very hot, being 200 times hotter than the visible surface of the Sun at 1 million °C.

Cosmology The study of the structure, evolution and fate of the Universe as a whole.

Crater A bowl-shaped depression on the surface of a planet, moon or asteroid formed when a meteorite or other asteroid hits the surface.

Crust The thin rocky outer surface of a planet or moon.

Day Length of time it takes for a planet to completely rotate on its axis. Earth's day is 23 hours 56 minutes and 4 seconds long.

Declination One of the two direction coordinates used to pinpoint an object in the sky. It tells us how far above or below the celestial equator a body lies, in degrees. It is the equivalent of latitude on Earth. The other coordinate is right ascension.

Doppler effect When an object is moving towards or away from us, its light changes colour. This is because the wavelengths of its light appear to shorten or lengthen. A police siren changes its pitch in the same way as it passes in the street.

Double star Two stars that look like one. These can be actually close together – binary stars. Or simply lined up by chance – optical pairs.

Dwarf planet A rounded body orbiting the Sun that is bigger than an asteroid but smaller than a planet. Pluto, Ceres and Eris are examples of dwarf planets.

Eclipse When one object passes in front of another. A solar eclipse is when the Moon obscures the Sun. Total solar eclipses can last for over seven minutes.

Ecliptic The apparent path of the Sun among the stars.

Electromagnetic radiation The radiation made of waves passing through changing electric and magnetic fields. Light is electromagnetic radiation, as are X-rays, ultraviolet light, infrared light and radio waves. The whole range of this radiation is called the electromagnetic spectrum.

Elliptical galaxy A type of galaxy that is oval-shaped. Elliptical galaxies tend to be made of older and redder stars.

Equator An imaginary line running around the middle of a planet or other body.

Equator, celestial The projection of the Earth's equator out into space. This forms a circle on the celestial sphere.

Equatorial mount A telescope stand that has a tilted axis. By making the tilt lie parallel to the axis of the Earth, you can keep a star or other object in view with a simple turning movement.

Equinox The two intersections of the ecliptic and the celestial equator. Day and night are of the same length at the equinoxes. The vernal equinox is around 21 March and the autumnal equinox around 22 September.

ESA The European Space Agency. This organisation has 20 member states and runs a large spaceflight program. It has its own spaceport in French Guiana.

Escape velocity The speed an object needs to reach to break free from the gravitational pull of a planet or other celestial body.

Extra-solar planet A planet that orbits a star other than our Sun. Also called an exoplanet. By early 2013, at least 862 exoplanets had been identified. Some of these may be good places to look for life.

Frequency For electromagnetic radiation, the number of waves that pass a point in one second.

Galaxy A system of stars, nebulae and interstellar matter, bound together by gravity.

Gamma ray Electromagnetic radiation with an extremely short wavelength. It is the most energetic form of radiation.

Gas giant A planet that doesn't have a solid surface, but is mostly made of gas and liquids with a small, dense core. Jupiter and Saturn are gas giants.

Gibbous phase The Moon's phase between half full and full.

Gravitational lensing When the gravity of a near object bends the light coming from an object further behind it. This can create multiple images of the far object.

Gravity The attractive force between objects. The higher the mass of an object, the greater its gravitational pull on other bodies.

Great Red Spot A huge storm on Jupiter, larger than the Earth, which has been raging for at least 300 years.

Heliopause The boundary where the solar wind flowing out from the Sun merges with the interstellar medium.

Heliosphere The area of space around the Sun where its influence is dominant. It is like a huge bubble of charged particles surrounding the Solar System.

Hertzsprung-Russell diagram (H-R diagram) This scatter graph plots stars according to their brightness and their temperature.

Hubble, Edwin An American astronomer who showed that there are more galaxies beyond our own. He also proved that the Universe is expanding.

Hubble Space Telescope A large telescope carried up into Earth orbit by the space shuttle in 1990. Hubble has made many amazing discoveries, including that there are probably black holes at the centre of every galaxy.

Ice giant A type of gas giant that has less hydrogen and more of the heavier elements such as oxygen, carbon, nitrogen, and sulphur. Uranus and Neptune are ice giants.

Inferior planets Mercury and Venus, the planets that are nearer to the Sun than the Earth.

Infrared Electromagnetic radiation with a wavelength just longer than that of red visible light. Every object that has a temperature emits infrared, letting astronomers see amazing sights that are hidden to the naked eye.

Inner planets The four planets closest to the Sun: Mercury, Venus, Earth and Mars.

Interstellar medium The sparse gas and dust that exists in the space between galaxies.

Irregular galaxy A type of galaxy without a symmetrical shape, that is neither spiral nor elliptical.

Kelvin scale A way of measuring temperature. 1K is equal to 1°C, but the Kelvin scale starts at absolute zero (-273.16°C). This is the coldest that anything can be.

Kuiper belt (also Edgeworth-Kuiper belt) An area beyond Neptune and Pluto where large numbers of comets orbit the Sun.

Lava flow Moving molten rock expelled from a volcano. Only Earth and the moons Io, Enceladus, Triton, and Europa currently have active volcanoes in our Solar System.

Light-year The distance a beam of light would travel in a vacuum in one year. One light-year is 9,460 billion km.

Luminosity The total energy a star emits in one second. This is a measure of its brightness.

Magnetosphere The zone of space around an object that is affected by the object's magnetic field.

Magnitude (apparent) This is a measure of an object's brightness as seen by an observer on Earth. The smaller the number, the brighter the object.

Mantle The area surrounding the rocky core of a planet. In the Earth, this lies just below the crust.

Mare (plural *maria*) A dark, smooth area on the surface of the Moon. Maria don't have many craters.

Messier number A catalogue of nebulae, star clusters and galaxies listed by Charles Messier in the late 18th century. There are 110 Messier objects.

Meteor Another term for a shooting star. A meteor is the bright streak of light caused when a small piece of space material enters the Earth's atmosphere and burns up.

Meteorite Any piece of a meteor that survives its descent into the atmosphere and lands on the ground. Meteorites are often metallic rocks.

Meteoroid A small piece of space debris, ranging from grain of sand- to boulder-sized. When meteoroids hit our atmosphere they become meteors.

Meteor shower A blizzard of meteors, when up to 1,000 can be seen every hour. They will seem to come, or radiate, from one point of the sky. A shower is caused by a cloud of space debris hitting our atmosphere.

Milky Way The spiral galaxy in which our Solar System is located. It contains over 200 billion stars and creates a broad band of hazy light in the night sky.

Moon A mini-planet that orbits another planet. Moons vary greatly in size and number. Earth has one moon, some planets have none, Jupiter has sixty-seven.

NASA The National Aeronautics and Space Administration. The part of the United States government that looks after space flights and exploration.

Nebula A cloud of dust and gas in space. There are different types of nebula, including: emission, reflection, dark, planetary and supernova remnant.

Neutron star The collapsed remains of a massive star that exploded as a supernova. Neutron stars are very hot and as dense as a Boeing 747 plane compressed to the size of a small grain of sand.

Newtonian reflector A type of telescope invented by Isaac Newton. It collects light from stars with a large mirror at the bottom of the telescope's tube. The light is then reflecting back up the tube to a smaller angled mirror and from there through an eyepiece to the observer.

NGC (New General Catalogue) A catalogue of 7,840 nebulae and star clusters compiled in 1888.

Nova An exploding star. This is usually caused by a white dwarf drawing in extra hydrogen from another star nearby. This causes a runaway nuclear reaction and the star flares up to many times its normal brilliancy before fading.

Oblate When a sphere isn't perfectly round, but seems to bulge slightly at its equator. Planets are often oblate because of the centrifugal force of their rotation.

Occultation When one celestial object covers up another.

Oort cloud A huge reservoir of comets that astronomers think surrounds the whole Solar System. It lies around one light-year from the Sun.

Orbit The path of a celestial object as it revolves around another body. For example, Earth orbits the Sun, the Moon orbits Earth.

Outer planets The four planets furthest from the Sun in our Solar System, beyond the asteroid belt. They are Jupiter, Saturn, Uranus and Neptune.

Phases The regular changes in the Moon's appearance. As the Moon orbits the Earth in a 27.5 day cycle, it moves relative to us and the Sun. We see different amounts of sunlight reflected from it. The pattern of phases moves from new moon to full moon and back again. The inferior planets also show phases.

Photosphere The bright, visible surface of the Sun.

Planet A body orbiting a star, that is big enough to have rounded itself with its own gravity, and to have cleared away nearby planetesimals.

Planetary nebula An expanding shell of glowing gas thrown off by a dying red giant star. The name was coined by an 18th century astronomer who thought this type of nebula looked like Uranus. Now also known as a stellar emission nebula.

Planetesimal Dust grains colliding in space often stick to each other, forming ever-growing lumps. When these lumps are about 1 km across, they have enough gravity to attract each other – they are planetesimals. When planetesimals combine they form protoplanets.

Plates The Earth's solid surface is broken up into moving pieces or plates. It is the only planet in the Solar System to have active plates.

Pole Planets have two poles, which are the furthest points away from the equator. They lie at either end of the axis on which the planet spins.

Poles, celestial Two imaginary points in the sky where the Earth's north and south poles would be if they were extended up into space.

Precession The slow, circling movement of the Earth's axis. This makes the celestial poles appear to move. It is caused by the tug of the Sun and Moon on the Earth's bulging equator. One complete cycle takes 26,000 years.

Prime meridian The line of longitude at 0°. This divides the Earth into East and West hemispheres. It is an imaginary line, and has been positioned in many places. It now runs through the Greenwich Observatory in London.

GLOSSARY

Prominences Huge plumes of glowing gas rising from the surface of the Sun. They often form a loop and can be 800,000 km long, half as wide as the Sun itself.

Protoplanet A baby planet, formed from kilometre-sized planetesimals gathering together under gravity. Several protoplanets can then attract each other and combine to form planets.

Protostar An early phase in the formation of a star. It forms when a huge gas cloud in the interstellar medium is hit by a shockwave and quickly contracts.

Pulsar A type of neutron star that rotates very rapidly and emits radio beams. These sweep round the cosmos like lighthouse beams, appearing as a regular pulse. The word is a contraction of 'pulsating star'.

Quasar The small, very bright core of a distant galaxy. Quasars are powered by material falling into black holes and are the brightest objects in the Universe. They can emit as much energy as several thousand normal galaxies.

Radiant The point in the sky where meteor showers seem to come from.

Red giant A star that has used up its hydrogen fuel and ballooned in size to several hundred times the size of our Sun. As it expands it cools and so appears more orange or red.

Reflecting telescope A telescope that uses mirrors to gather light. Reflecting telescopes can be larger and more powerful than refractors, but they are more expensive.

Refracting telescope A telescope that uses lenses to gather light. A refracting telescope can be made relatively cheaply, but its size is limited.

Relativity A theory proposed by Albert Einstein. Among other things it states that energy and mass are equivalent and that nothing can exceed the speed of light. It revolutionized physics and astronomy and helped predict black holes and neutron stars.

Retrograde motion When an object orbits or rotates in the opposite direction to another body. Venus spins the opposite way to Earth, so its rotation is retrograde.

Right ascension One of the two direction coordinates used to pinpoint an object in the sky. It tells us how far round the celestial equator a body lies, in degrees. It is the equivalent of longitude on Earth. The other coordinate is declination.

Rotate Planets spin around a central axis that runs from pole to pole. This axis is often tilted a little. The Earth rotates around its axis once every 23 hours 56 minutes and 4 seconds.

Satellite Any object in outer space that goes around another object. For example, the Moon is a satellite of Earth, as is the Hubble Space telescope.

Scintillation The twinkling of a star. This effect is caused by Earth's atmosphere.

Silicate A compound containing the element silicon.

Solar 'Sol' is a Latin word that translates as Sun. So 'solar' means anything to do with the Sun.

Solar flare A sudden bright surge of energy from the surface of the Sun. Flares are extremely powerful and hot, reaching tens of millions of degrees Kelvin. They can disrupt radio communications on Earth.

Solar nebula A cloud of interstellar dust and gas from which the Solar System formed around 4.6 billion years ago.

Solar System The Sun and all the bodies orbiting it, as well as other celestial objects within its gravitational field. This includes the eight planets, their moons, dwarf planets and billions of other space objects such as asteroids, comets and dust particles.

Solar wind A flow of charged particles streaming out constantly from the Sun's upper atmosphere in all directions.

Solstice The two times in the year when the Sun is furthest from the celestial equator. In each hemisphere this corresponds to the longest summer day and the shortest winter day.

Spiral galaxy A galaxy that is roughly circular in shape, within which the stars, gas and dust sweep in long spiral arms out from the centre.

Star A ball of exploding gas held together by its own gravity.

Subduction When one tectonic plate moves under another into the Earth's mantle.

Sun The yellow star at the centre of our Solar System. Medium-sized as stars go, it has a diameter 100 times that of Earth. It formed 4.6 billion years ago.

Supercluster A gathering of several groups and clusters of galaxies. Superclusters are the largest known structure in the Universe and may be up to a billion light-years across.

Superior planets The planets beyond Earth in the Solar System: Mars, Jupiter, Saturn, Uranus, Neptune.

Supernova An exploding star that is even more powerful than a nova. The burst of energy can outshine an entire galaxy before fading after a few weeks.

Tail A tail, or coma, is the visible dust and gas that melts off a comet as it passes by the Sun. Comet tails are blown away from the Sun by the solar wind.

Terrestrial planet Planets made up mostly of rock, such as the four planets of the inner Solar System (Mercury, Venus, Earth and Mars).

Transit The movement of a small body across the disk of a larger body. Mercury and Venus sometimes transit the Sun.

Ultraviolet light A type of electromagnetic radiation. Most stars emit some ultraviolet radiation – ten per cent of sunlight's energy is ultraviolet. Hotter objects emit more ultraviolet, so astronomers use it to make images of young massive stars and very old stars and galaxies. These are hotter near their birth or death.

X-rays A type of electromagnetic radiation, high in energy. Many space objects emit them, including exploded stars. Since Earth's atmosphere absorbs X-rays, detectors work best when lifted high above the planet's surface, like the Chandra Observatory.

Waning Moon The phase when the Moon's sunlit portion is getting smaller, from full to new.

Waxing The phase when the Moon's sunlit portion is getting bigger, from new to full.

White dwarf The small, very dense star that remains when a star like our Sun has used up its nuclear fuel. About one in ten of the stars in the Milky Way are white dwarfs.

Year The length of time it takes a planet to go around the Sun. Earth's year is 365 days and 6 hours long.

Zenith The point in the night sky directly over your head.

Zodiac The belt that straddles 8° to either side of the ecliptic and stretches round the night sky. The Sun, Moon and principal planets stay within the zodiac.

Zodiacal light A faint cone of light in the night sky. It stretches from the horizon up along the ecliptic and is caused by sunlight scattering in space dust. It is best seen just after sunset and before sunrise in spring and autumn.

NDEX

INDEX

ACKNOWLEDGEMENTS

Main text: Richard Happer

Photo Credits

Front cover: Enrico Agostoni/Shutterstock

Back cover: Saturn, NASA/JPL; Spirograph Nebula, NASA and The Hubble Heritage Team (STScI/AURA)/ R. Sahai et al; Moon and Venus, Grant Glendinning/Shutterstock; M83, ESO.

Background images: Blue Stars, pixelparticle/Shutterstock; Milky Way, inigo cia/Shutterstock; Purple nebula, Anatolii Vasilev/Shutterstock; Stars above Malta, McCarthy's Photoworks/Shutterstock; tycho8, Data source: Eros. Created by Seal. From the database of JPL/Caltech generated planetary maps. Author/Origin: NASA Jet Propulsion Laboratory – Solar System Simulator http://space.jpl.nasa.gov/; Phoenix, Ysbrand Cosijn/Shutterstock; Night sky, McCarthy's Photoworks/Shutterstock.

Star charts: © IAU/Sky and Telescope (IAU with Sky & Telescope magazine (Roger Sinnott & Rick Fienberg) and Alan MacRobert).

14, Eros, NASA/Goddard Space Flight Center Scientific Visualization Studio; 15, Hale-Bopp, NASA; 15, Meteroid, max voran/Shutterstock; 20, Milky Way illustration, Vividfour/Shutterstock; 20, Sun spot, John R Smith/Shutterstock; 21, Sun flare, NASA; 24, Mercury, NASA/Johns Hopkins University Applied Physics Laboratory/Carnegie Institution of Washington; 25, Ceres, NASA; 30, Caloris Basin, NASA; 31, Mercury transit, ©Phil Jones; 36, Venus clouds, NASA; 37, Gula Mons, NASA/JPL; 45, Mt Etna, Fredy Thuerig/Shutterstock; 47, Atmosphere, NASA/JPL/UCSD/JSC; 50, Lunar eclipse, Primo_Cigler/Shutterstock; 50, Solar eclipse, Vladimir Wrangel/Shutterstock; 52, By the crater, NASA; 52, Lunar lander, NASA; 53, Far side of the Moon, NASA/Goddard/Arizona State University; 58, Mars gullies, NASA/JPL-Caltech/University of Arizona; 59, Mars from Pathfinder, NASA/JPL; 59, Phobos, G. Neukum (FU Berlin) et al. Mars Express, DLR, ESA; 62, Asteroids, NASA/JPL-Caltech/JAXA/ESA; 63, Ida nine views, NASA/JPL/USGS; 74, Io surface, NASA/JPL; 75, Galilean moons, NASA/JPL-Caltech; 75, Jupiter and moons, © Jan Sandberg/desert-astro. com; 75, Thebe, NASA/JPL; 76 Jupiter, Shutterstock; 76, Jupiter interior, G. Neukum/NASA; 77, Great Red Spot, NASA/JPL; 80, Great White Spot, NASA/JPL-Caltech/SSI; 82, Saturns rings tilt, NASA and The Hubble Heritage Team (STScI/AURA); 83, Coloured rings, NASA; 83, Saturn edge on, Erich Karkoschka (University of Arizona Lunar & Planetary Lab) and NASA; 86, Titan behind rings, NASA/JPL/Space Science Institute; 87, Saturn's moons, NASA/JPL; 88, Saturn storm, NASA/JPL; 89, Saturn from above, Gl0ck/Shutterstock; 94, Uranus moons, ©Vzb83 based on NASA images; 94–95, Uranus, Shutterstock; 95, Miranda, NASA/JPL; 100, Neptune, Shutterstock; 100, Neptune clouds, NASA/JPL; 101, Neptune Great Dark Spot, NASA; 101, Triton, NASA/JPL; 108, Parts of a comet, Mark R/Shutterstock; 109, Comet McNaught, ESO; 109, Tempel 1, NASA/JPL-Caltech/UMD; 111, Pluto, NASA; 116, Planetary nebula, NASA and The Hubble Heritage Team (STScI/AURA); 116, Star forming nebula, Giovanni Benintende/Shutterstock; 116–117, Sun illustrations, Zhabska Tetyana/Shutterstock; 117, Black hole, NASA; 117, Neutron star, NASA; 117, NGC 3603, NASA, Wolfgang Brandner (JPL/IPAC), Eva K. Grebel (Univ. Washington), You-Hua Chu (Univ. Illinois Urbana-Champaign); 117, Supernova, Filatov Alexey/Shutterstock; 120, Cat's Eye, X-ray: NASA/CXC/SAO; Optical: NASA/STScI; 120, Spirograph Nebula, NASA and The Hubble Heritage Team (STScI/AURA)/ R. Sahai et al.; 121, Cassiopeia A, NASA/JPL-Caltech/STScI/CXC/SAO; 121, Crab Nebula, NASA, ESA, J. Hester (Arizona State University); 121, Barnard 68, FORS Team, 8.2-meter VLT Antu. ESO; 121, LBN114.55+00.22, NASA/JPL-Caltech/UCLA; 121, Hen 3-1475, ESA/NASA; 121, Witch Head Nebula, NASA STScI Digitized Sky Survey/Noel Carboni; 123, Hubble fork images, Messier.seds.org; 124, Hubble XDF, NASA, ESA, G. Illingworth, D. Mageeand P. Oesch (UCSC), R. Bowens (Leiden Obs.) and the XDF Team; 124, M101, X-ray: NASA/CXC/SAO; IR & UV: NASA/JPL-Caltech; Optical: NASA/STScI; 125, NGC 1132, NASA and The Hubble Heritage Team (STScI/AURA)-ESA/Hubble Collaboration/ M. West (ESO, Chile); 125, Interacting galaxies, ESO; 125, NGC 3115, X-ray: NASA/CXC/Univ. of Alabama/K. Wong et al., Optical: NASA/STScI; 125, NGC 1300, NASA, ESA, and The Hubble Heritage Team (STScI/AURA); 125, Small Magellanic Cloud, ESA/NASA/JPL-Caltech/STScI; 130, M31, ESO/S. Brunier; 131, M31, NASA/JPL-Caltech; 131, Arp 273, NASA, ESA, and Hubble Heritage Team (STScI/AURA); 131, NGC 7662, X-ray: NASA/CXC/RIT/J. Kastner et al.; Optical: NASA/STScI; 132, NGC 7009, X-ray: NASA/CXC/RIT/J. Kastner et al.; Optical: NASA/STScI; 133, Atoms-for-Peace, ESO; 133, Lyman Alpha blobs, X-ray (NASA/CXC/Durham Univ./D. Alexander et al.); Optical (NASA/ESA/STScI/IoA/S. Chapman et al.); Lyman-alpha Optical (NAOJ/Subaru/Tohoku Univ./T. Hayashino et al.); Infrared (NASA/JPL-Caltech/Durham Univ./J. Geach et al.)); 133, M72 cluster, NASA, ESA, Hubble, HPOW; 134, Boötes field, X-ray: NASA/CXC/CfA/R. Hickox et al.; 135, Chandra illustration and Cloverleaf Quasar, X-ray: NASA/CXC/Penn State/G. Chartas et al; Illustration: NASA/CXC/M. Weiss; 135, GB 1428+4217, X-ray: NASA/CXC/SAO/J. Drake et al., Optical: Univ. of Hertfordshire/INT/IPHAS, Infrared: NASA/JPL-Caltech; 135, VV340, X-ray NASA/CXC/IfA/D. Sanders et al; Optical NASA/STScI/NRAO/A. Evans et al.; Infrared and Ultraviolet NASA/JPL-Caltech/J. Mazzarella et al.; 136, Beehive cluster, Atlas Image courtesy of 2MASS/UMass/IPAC-Caltech/NASA/NSF; 137, Abell 30, X-ray (NASA/CXC/IAA-CSIC/M. Guerrero et al); Optical (NASA/STScI); 137, Musket Ball cluster, X-ray: NASA/CXC/UCDavis/W. Dawson et al; Optical: NASA/STScI/UCDavis/W. Dawson et al.; 138, Sirius, ESO/Y. Beletsky; 138, Sirius A and B, NASA/SAO/CXC; 141, M103, Hillary Mathis, N. A. Sharp/AURA/NOAO/NSF; 141, Tycho supernova remnant, MPIA/NASA/Calar Alto Observatory; 141, W3, X-ray: NASA/CXC/Penn State/L. Townsley et al. Optica:l Pal Obs. DSS142; 143, Proxima Centauri, Atlas Image courtesy of 2MASS/UMass/IPAC-Caltech/NASA/NSF; 143, Alpha Centauri, ESO/Digitized Sky Survey 2/Davide De Martin; 143, Centaurus A, X-ray: NASA/CXC/CfA/R. Kraft et al.; Submillimeter: MPIfR/ESO/APEX/A. Weiss et al.; Optical: ESO/WFI; 143, NGC 4622, NASA and The Hubble Heritage Team (STScI/AURA)/Dr Ron Buta et al.; 145, Jewel Box, ESO/Y. Beletsky; 145, Reflection nebula, NASA/JPL-Caltech/UCLA; 145, Southern Cross, ESO/Y. Beletsky; 147, Castor and Pollux, ©j-dub1980147; Eskimo Nebula, NASA, Andrew Fruchter and the ERO Team [Sylvia Baggett (STScI), Richard Hook (ST-ECF), Zoltan Levay (STScI)]; 147, Geminids, ESO/G. Lombardi (glphoto.it); 147, M35 and NGC 2158, Atlas Image courtesy of 2MASS/

UMass/IPAC-Caltech/NASA/NSF; 148, Hydra A, NASA/CXC/SAO Radio: NRAO; 149, Ghost of Jupiter, Bruce Balick and Jason Alexander (University of Washington), Arsen Hajian (U.S. Naval Observatory), Yervant Terzian (Cornell University), Mario Perinotto (University of Florence), Patrizio Patriarchi (Arcetri Observatory) and NASA/ESA; 149, M83, ESO; 149, M83 (part), NASA and The Hubble Heritage Team (STScI/AURA); 149, Overlapping Galaxies, NASA and The Hubble Heritage Team (STScI/AURA); 150, Betelgeuse trio, ESO, P. Kervella, Digitized Sky Survey 2 and A. Fujii; 151, Barnard's Loop, perezanz/Shutterstock; 151, Orion Nebula, Digitized Sky Survey, ESA/GEO/NASA FITS Liberator Color Composite: Davide De Martin (Skyfactory); 152, Stephan's Quintet, NASA, ESA, and the Hubble SM4 ERO Team; 153, Einstein Cross, ESA/Hubble & NASA; 153, Interacting galaxies, NASA, ESA, and The Hubble Heritage Team (STScI/AURA)-ESA/Hubble Collaboration and A. Evans et al.; 153, M15, ESA, Hubble, NASA; 153, M15 Nuetron Star, NASA/GSFC/N. White, L. Angelini; 155, Perseids, John A Davis/Shutterstock; 155, Perseus A, X-ray: NASA/CXC/IoA/A. Fabian et al.; Radio: NRAO/VLA/G. Taylor; Optical: NASA/ESA/Hubble Heritage (STScI/AURA) & Univ. of Cambridge/IoA/A. Fabian; 155, Perseus Galaxy cluster, NASA, ESA, and the Digitized Sky Survey 2 Davide De Martin (ESA/Hubble); 155, Pulsar, NASA/CXC/IUSS/A. De Luca et al.; 157, Arp 227, NASA/JPL-Caltech/L. Lanz (Harvard-Smithsonian CfA); 157, M74, NASA, ESA, and The Hubble Heritage (STScI/AURA)-ESA/Hubble Collaboration; 157, NGC 520, NASA ESA, the Hubble Heritage Team (STScI/AURA)-ESA/Hubble Collaboration and B. Whitmore (STScI).; 159, Galactic Centre, NASA & ESA; 159, M8, A. Caulet (ST-ECF, ESA) and NASA; 159, Sagittarius A, NASA/Penn State/G. Garmire et al.; 159, The Mouse, NASA/CXC/SAO/B. Gaensler et al. Radio: NSF/NRAO/VLA; 161, Antares and M4, Giovanni Beinintende/Shutterstock; 161, Butterfly Nebula, NASA, ESA, and the Hubble SM4 ERO Team; 161, Cat's Paw Nebula, ESO/R. Gendler & R. M. Hannahoe; 162, Pleiades and Hyades, Valerio Pardi/Shutterstock; 163, Crab Nebula, X-ray: NASA/CXC/SAO/F. Seward; Optical: NASA/ESA/ASU/J. Hester & A. Loll; Infrared: NASA/JPL-Caltech/Univ. Minn./R. Gehrz; 163, Pleiades, NASA, ESA and AURA/Caltech; 163, Taurus Molecular Cloud, ESO/APEX (MPIfR/ESO/OSO)/A. Hacar et al./Digitized Sky Survey 2. Davide De Martin.; 165, Grey Bear, Nadezhda Bolotina/Shutterstock; 165, I Zwicky 18, NASA, ESA and A. Aloisi (Space Telescope Scince Institute and ESA, Baltimore, Md.); 165, Lockman Hole, ESA/Herschel/SPIRE/HerMES; 165, M81 and M82, John A Davis/Shutterstock; 166, Polaris, John A Davis/Shutterstock; 167, Comet Hyakutake, Andrew F Kazmierski/Shutterstock; 167, NGC 6217, NASA, ESA, and the Hubble SM4 ERO Team; 167, Star Trails, SeanPavonePhoto/Shutterstock; 168, M87, X-ray: NASA/CXC/KIPAC/N. Werner, E. Million et al; Radio: NRAO/AUI/NSF/F. Owen; 169, Arp 274, NASA, ESA, and M. Livio and the Hubble Heritage Team (STScI/AURA); 169, Sombrero Galaxy, ESO/ P. Barthel & (M. Neeser and R. Hook); 169, Virgo Cluster, NASA, ESA, M. Postman (STScI), and the CLASH Team; 172, Mercury transit, SOHO (ESA & NASA); 173, Conjunction, ESO/Y. Beletsky; 173, Cookie crater, NASA/Johns Hopkins University Applied Physics Laboratory/Carnegie Institution of Washington; 173, Happy crater, NASA; 174, Moon and Venus, Grant Glendinning/Shutterstock; 174, Venus transit, NASA/SDO & the AIA, EVE, and HMI teams; 175, Crater Farm, NASA/JPL; 176, Dust Devil, NASA/JPL-Caltech/Univ. of Arizona; 176, Mars ice cap, NASA, ESA, and The Hubble Heritage Team (STScI/AURA); 177, Rocknest, NASA/JPL-Caltech/Malin Space Science Systems, 178, Planets in sky, ESO/Max Alexander; 178, Spots, NASA, ESA, and A. Simon-Miller (NASA Goddard Space Flight Center); 179, Comet Marks, Hubble Space Comet Team and NASA; 180, Saturn close, NASA/JPL-Caltech/Space Science Institute; 180, Saturn distant, McCarthy's Photoworks/Shutterstock; 181, Moons transiting, NASA/JPL/STSI; 181, Titan surface, NASA/JPL-Caltech/USGS; 182, Apollo 11, NASA; 182, Astronaut, NASA; 182, Mars Curiosity Rover, NASA; 182, Sputnik 1 replica, NASA/NSSDC; 183, Hubble Telescope, NASA; 183, Space Shuttle, NASA; 183, Voyager 2, NASA/JPL; 184–185, all images, Image Science and Analysis Laboratory, NASA-Johnson Space Center. The Gateway to Astronaut Photography of Earth.

14, 18, 20, 24, 28, 30, 34, 36, 40, 46, 56, 58, 66, 70, 76, 80, 88, 92, 94, 95, 98, 100, 101, 104, 110, 175, Sun, Anthony McAulay/Shutterstock; 15, 109, comet, Shutterstock; 19, 37, 46, 47, Big Sun, xfox01/Shutterstock; 19, 21, 25, 29, 34, 37, 47, 51, 57, 72, 77, 81, 89, 93, 99, 105, 111, 175, Earth, NASA; 46, 50, Earth Blue Marble, NASA; 25, 57, 111, Mars, NASA/JPL; 25, 35, 37, 111, 175, Venus, NASA/JPL; 29, 31, 110, Mercury, Luis Stortini Sabor aka CVADRAT/Shutterstock; 14–15, 18–19, 24, 28, 30, 34, 36, 40, 46, 52, 56, 58, 66, 70, 80, 88, 92, 98, 104, Solar System, Shutterstock; 19, 67, 72, 89, 110, Jupiter, NASA/JPL/University of Arizona; 50, 51, 52, Moon, Lick Observatory; 67, 93, 101, 105, 111, Neptune, NASA; 67, 81, 89, 110, Saturn, George Toubalis/Shutterstock; 67, 93, 111, Uranus, NASA/Space Telescope Science Institute.

All other illustrations © CollinsBartholomew Ltd

Useful Websites

Astronomy Central	astronomycentral.co.uk/planets-to-see-in-thesky-tonight
Astronomy Today	www.astronomytoday.com/skyguide.html
Astro-Observer	astro-observer.com/index.html
Chandra	chandra.harvard.edu
ESA	www.esa.int/ESA
ESO	www.eso.org/public/
Herschel	www.herschel.caltech.edu
Hubble	hubblesite.org
IAU	www.iau.org
International Dark-Sky Association	www.darksky.org/night-sky-conservation/national-park-service
Messier	messier.seds.org
NASA	www.nasa.gov
NASA JPL	www.jpl.nasa.gov